快速掌握 PostgreSQL 版本新特性

主编 彭 冲

副主编 高云龙 阎书利 类延良

参编 吕新杰 白国华 魏 波 王其达 李孟洁

电子工业出版社
Publishing House of Electronics Industry
北京·BEIJING

内 容 简 介

自 2019 年底加入墨天轮数据社区以来，笔者在 PG 乐知乐享专栏持续撰写了与 PostgreSQL 相关的文章，近几年陆续撰写了一些有关各版本新特性方面的文章。由于 PostgreSQL 每个版本发布的新特性较多，因此要想全面、完整地掌握 PostgreSQL 新特性十分困难。本书对 PostgreSQL 的 7 个大版本：从 PostgreSQL 10 到 PostgreSQL 16，根据公开的新特性实验手册、新特性相关的文章、邮件列表、社区核心提交者相关的博客，以及官方 Release Notes 等素材进行精加工，并从主要性能、可靠性、运维管理、开发易用性、系统层 5 个方面进行介绍。在当前国内外环境下，PostgreSQL 正在不断崛起，本书从新特性的视角帮助读者掌握数据库的正确使用方式。

未经许可，不得以任何方式复制或抄袭本书之部分或全部内容。
版权所有，侵权必究。

图书在版编目（CIP）数据

快速掌握 PostgreSQL 版本新特性 / 彭冲主编. —北京：电子工业出版社，2023.12
ISBN 978-7-121-46740-0

Ⅰ. ①快… Ⅱ. ①彭… Ⅲ. ①关系数据库系统 Ⅳ. ① TP311.132.3

中国国家版本馆 CIP 数据核字（2023）第 223537 号

责任编辑：杜　军
印　　刷：河北虎彩印刷有限公司
装　　订：河北虎彩印刷有限公司
出版发行：电子工业出版社
　　　　　北京市海淀区万寿路 173 信箱　邮编：100036
开　　本：787×1092　1/16　印张：13.5　字数：337 千字
版　　次：2023 年 12 月第 1 版
印　　次：2025 年 3 月第 4 次印刷
定　　价：45.00 元

凡所购买电子工业出版社图书有缺损问题，请向购买书店调换。若书店售缺，请与本社发行部联系，联系及邮购电话：(010) 88254888，88258888。
质量投诉请发邮件至 zlts@phei.com.cn，盗版侵权举报请发邮件至 dbqq@phei.com.cn。
本书咨询联系方式：dujun@phei.com.cn。

序一

开源是数字经济公共基础设施，PostgreSQL作为世界上最优秀的开源数据库，是公认的产权公有的全球性技术。自80年代诞生以来，包括中国在内的全球市场中PostgreSQL占据关键生态位置。与Linux、MySQL等全球性开源基础技术一道，支撑着过去及未来全球信息技术和数字化的爆炸式发展，成为数字社会运行的"基础底座"。PostgreSQL全球开发者社区所倡导的自由、民主、泛众的开源精神，团结了世界各地的开源开发人才，推动PostgreSQL项目稳定迭代。成为全球性开源协同创新的典范。充实丰富了全球开源文化的思想内涵。

工信部中国开源软件推进联盟是我国开源领域最早成立的社会组织。基于PostgreSQL技术重要性，2017年联盟正式成立PostgreSQL分会，承担PostgreSQL行业协会组织的职能。致力于推动PostgreSQL技术在中国的传播与应用，社区的组织与运营，产业生态的建构与实践，和产学研用一体化发展水平的提升。

彭冲作为中国PostgreSQL分会的PostgreSQL ACE，经常活跃在PostgreSQL社区，为PGFans解答问题，为客户提供专业服务。本次笔者结合多年工作实践，面向PostgreSQL 10到16主版本，系统化阐述了其功能特性及示例。相比于PostgreSQL特性技术文档的分散、难以理解和缺少示例，通过本书能够极大的帮助PostgreSQL用户节约时间，快速便捷地理解和使用PostgreSQL功能特性。对作者能够潜心面向PostgreSQL 7个主版本、200多个技术特性亲身整理、实验验证，表示赞许。本书作为PostgreSQL培训认证的指定教材，将助力认证学员更高效地学习、掌握PostgreSQL技术，有助于推动PostgreSQL在中国的生态发展。借此也向传承开源精神的每位PostgreSQL贡献者致以敬意。

中科院软件所党委委员、时空数据管理与数据科学研究中心主任、中国开源软件推进
联盟PostgreSQL分会主席　丁治明

序二

PostgreSQL是一款功能非常强大的数据库，根据Stack Overflow的开发者调查，如今PostgreSQL已超越MySQL成为最受欢迎的开源数据库。从DB-Engines网站也观察到，PostgreSQL已经四次成为年度数据库，分别是2017年、2018年、2020年、2023年。PostgreSQL强大的功能离不开PostgreSQL全球开发组坚持每年的大版本更新，每次大版本发布提供的新功能都会让开发者眼前一亮。

随着PostgreSQL的功能越来越丰富，开发者对数据库的功能、特征进行全面掌握越来越有难度，使得PostgreSQL就像一个高级武器一样，难以发挥出其功力来为业务降本提效，因此，它还有巨大的潜力可以挖掘。

彭冲编写的《快速掌握PostgreSQL版本新特性》特别适合开发者、架构师阅读，通过了解PostgreSQL的新特性，相信我们的开发者和架构师可以更好地发挥数据库的潜能，助力业务创新发展。

PostgreSQL社区布道师　德哥

前 言
Preface

1．为什么要写这本书

PostgreSQL 官方文档及 Release Notes 对新特性的介绍比较晦涩难懂，国内外一些"大佬"的博客和历届 PostgreSQL 大会也都有各版本新特性的详细介绍，但都比较分散。另外，PostgreSQL 全球社区发布新版本的速度十分快，大多数人都来不及学习每个版本的新特性。笔者多年前入驻墨天轮数据社区，受乐知乐享的精神鼓励，在该平台创作了有关 PostgreSQL 的文章一百多篇。得益于 PostgreSQL 相关社区平台及技术交流微信群，笔者输出了大量干货文章，其中包括很多版本新特性的实践案例，并将其整理成书稿。

2．本书的主要内容

本书对国内外有关 PostgreSQL 新特性相关的文章进行归纳整理，从 PostgreSQL 10 到 PostgreSQL 16，共分 7 章，从主要性能、可靠性、运维管理、开发易用性、系统层 5 个方面进行介绍，同时基于各版本最新发布的小版本进行实践验证，并对其加工修订。本书中的上百个新特性均有实际案例作为支撑。如果您对下面这些问题感兴趣，那么学习本书中的知识会对您有所帮助。

- PostgreSQL 有哪些兼容 Oracle 的功能特性？

PostgreSQL 11 支持事务控制的存储过程，PostgreSQL 12 支持绑定变量窥探设置优化器策略，PostgreSQL 14 支持查询 ID 分析慢查询，PostgreSQL 15 支持 MERGE 语句，PostgreSQL 16 支持子查询不带别名。

- PostgreSQL 有哪些性能得以提升？

PostgreSQL 10 的数据分区，PostgreSQL 11 的 B-Tree 索引并行创建，PostgreSQL 13 的索引去重，PostgreSQL 14 的高并发连接优化，PostgreSQL 15 的统计信息内存化，PostgreSQL 16 的 VACUUM 操作使用的 ring buffer 内存可控。

- PostgreSQL 有哪些易用的开发函数？

PostgreSQL 10 的 xmltable 函数，PostgreSQL 13 的 UUID 数据类型的内置函数，PostgreSQL 14 的 string_to_table 函数，PostgreSQL 16 的 any_value 函数等。

- PostgreSQL 的逻辑复制有哪些改进？

PostgreSQL 11 支持 TRUNCATE 操作，PostgreSQL 13 支持分区表同步及同步到异构分区表，PostgreSQL 14 支持流式处理大事务、提升 TRUNCATE 操作的性能、支持以二进制形式传输数据及优化表的初始数据同步等，PostgreSQL 15 支持两阶段提交、允许发布模式下的所有表、支持行级和列级过滤及改进复制冲突处理，PostgreSQL 16 支持从 standby 节点复制、发布端并行应用大事务。

3．适合阅读本书的读者

- 所有对 PostgreSQL 感兴趣的人。

本书可以帮助不太熟悉 PostgreSQL 的 DBA（数据库管理员）通过运维管理的新特性实践快速成长为一名合格的 PostgreSQL DBA。PostgreSQL 爱好者通过系统学习本书各版本新特性的知识也可以成为 PostgreSQL 专家。熟悉 PostgreSQL 的 DBA 还可以利用各版本新特性达到事半功倍的效果，进一步提高自身的数据库运维管理水平。

- 国产数据库 DBA。

本书也非常适合转型到基于 PostgreSQL 内核的国产数据库 DBA，通过新特性的版本演进可以掌握国产数据库的特性。

- 开发人员。

开发人员可以通过本书开发相关的新特性掌握与 PostgreSQL 相关的标准语法和内置函数等的技巧，提高开发效率。

PostgreSQL 正在经历一场缓慢的崛起，读者可以考虑投入更多的关注与学习。

4．勘误和支持

由于笔者的水平有限，书中难免存在一些疏漏，敬请读者批评指正。读者可以将在书中遇到的问题及宝贵意见发送至 chong.peng@enmotech.com，或者添加笔者的微信号 skypkmoon，笔者很期待听到您的真挚反馈。

5．致谢

感谢 2015—2018 届 PostgreSQL 中文社区主席萧少聪及现任主席张文升带领笔者进入 PostgreSQL 的大门；感谢 PostgreSQL 中文社区大学校长周正中（德哥）日复一日持续的公益奉献；感谢知名微信公众号"PostgreSQL 学徒"的作者熊灿灿和"非法加冯"的作者冯若航对相关案例的分享；感谢在 PGFans 技术社区和墨天轮数据社区大量发布关于 PostgreSQL 新特性文章的作者，让笔者可以学习借鉴；感谢中国 PostgreSQL 分会的大力支持，感谢电子工业出版社的编辑杜军老师对本书出版的大力支持；最后感谢笔者的妻子及母亲，她们的默默付出让笔者能坚持把这本书写完。

2023 年 10 月

目 录

第 1 章 PostgreSQL 10 新特性 1

1.1 PostgreSQL 10 的主要性能提升 1
  1.1.1 数据分区 1
  1.1.2 并行特性增强 1
  1.1.3 统计信息扩展 2
  1.1.4 AFTER 触发器性能增强 4
  1.1.5 Aggregate 操作支持下推 5

1.2 PostgreSQL 10 的可靠性提高 7
  1.2.1 仲裁提交引入 7
  1.2.2 HASH 索引支持记录 WAL 8
  1.2.3 事务提交状态检测 8

1.3 PostgreSQL 10 的运维管理优化 9
  1.3.1 WAL 文件大小扩容 9
  1.3.2 WAL 支持在线压缩 10
  1.3.3 ICU 标准库引入 10
  1.3.4 活动会话视图增强 10
  1.3.5 HBA 文件新增系统视图 11
  1.3.6 promote 子命令增加等待模式 12
  1.3.7 基础备份增强 12
  1.3.8 临时复制槽引入 13
  1.3.9 行级安全策略优化 13
  1.3.10 PSQL 工具允许条件式交互 15

1.4 PostgreSQL 10 的开发易用性提升 15
  1.4.1 声明式分区引入 15
  1.4.2 表级数据发布与订阅 17
  1.4.3 标识列引入 18
  1.4.4 全文检索支持 JSON 与 JSONB 数据类型 21
  1.4.5 xmltable 函数引入 22

1.5 PostgreSQL 10 的系统层变化 ... 24
1.5.1 XLOG 重命名 ... 24
1.5.2 系统元数据引入 ... 25
1.5.3 配置参数引入 ... 26
1.5.4 口令加密安全性提高 ... 26
1.5.5 预置角色变化 ... 26
1.5.6 附加模块变化 ... 27
1.6 本章小结 ... 27

第 2 章 PostgreSQL 11 新特性 ... 28
2.1 PostgreSQL 11 的主要性能提升 ... 28
2.1.1 WAL 可配置 ... 28
2.1.2 B-Tree 索引并行创建 ... 29
2.1.3 HASH 操作及 HASH JOIN 操作支持并行 ... 30
2.1.4 其他并行特性支持 ... 31
2.1.5 表达式索引引入 ... 32
2.1.6 覆盖索引引入 ... 33
2.1.7 实时编译引入 ... 34
2.1.8 缓存管理改进 ... 35
2.1.9 UPDATE 操作和 DELETE 操作支持下推 ... 35
2.2 PostgreSQL 11 的可靠性提高 ... 36
2.2.1 数据块校验和检测 ... 36
2.2.2 B-Tree 索引坏块检测 ... 37
2.2.3 查询 ID 由 32 位扩充为 64 位 ... 37
2.3 PostgreSQL 11 的运维管理优化 ... 38
2.3.1 快速添加列 ... 38
2.3.2 维护操作支持多个表 ... 40
2.3.3 分区数据支持通过父表加载 ... 40
2.3.4 新增 KILL 信号 ... 40
2.3.5 WAL 支持离线重构 ... 40
2.3.6 PSQL 工具支持记录语句执行情况 ... 41
2.4 PostgreSQL 11 的开发易用性提升 ... 42
2.4.1 声明式分区增强 ... 42
2.4.2 支持事务控制的存储过程 ... 43
2.4.3 逻辑复制支持 TRUNCATE 操作 ... 44
2.4.4 窗口函数增强 ... 44
2.5 PostgreSQL 11 的系统层变化 ... 45
2.5.1 配置参数引入 ... 45

2.5.2 预置角色变化 ... 46
2.5.3 超级用户权限下放 ... 46
2.5.4 附加模块变化 ... 47
2.6 本章小结 ... 47

第3章 PostgreSQL 12 新特性 ... 48

3.1 PostgreSQL 12 的主要性能提升 ... 48
   3.1.1 CTE 优化 ... 48
   3.1.2 索引效率提升 ... 50
   3.1.3 系统函数优化 ... 51
3.2 PostgreSQL 12 的运维管理优化 ... 51
   3.2.1 校验和开关控制 ... 51
   3.2.2 COPY FROM 命令数据过滤 ... 52
   3.2.3 用户级流复制超时控制 ... 52
   3.2.4 VACUUM 操作及 ANALYZE 操作锁跳过 ... 53
   3.2.5 表及索引清理解耦 ... 54
   3.2.6 索引在线重建 ... 54
   3.2.7 执行计划显示非默认参数 ... 55
   3.2.8 后台操作进度报告引入 ... 56
   3.2.9 备库升主库开放 SQL 接口 ... 57
   3.2.10 PSQL 工具帮助链接添加 ... 57
3.3 PostgreSQL 12 的开发易用性提升 ... 58
   3.3.1 声明式分区增强 ... 58
   3.3.2 运算存储列使用 ... 59
   3.3.3 绑定变量窥探引入 ... 60
   3.3.4 SQL/JSON path 引入 ... 61
   3.3.5 枚举数据类型增强 ... 63
3.4 PostgreSQL 12 的系统层变化 ... 63
   3.4.1 表存储引擎开放 ... 63
   3.4.2 恢复相关配置优化 ... 64
   3.4.3 系统元数据引入 ... 64
   3.4.4 配置参数引入 ... 65
   3.4.5 流复制连接数优化 ... 66
   3.4.6 DOS 攻击预防 ... 66
   3.4.7 SSL 协议可控 ... 67
   3.4.8 附加模块变化 ... 67
3.5 本章小结 ... 67

第 4 章 PostgreSQL 13 新特性 68

4.1 PostgreSQL 13 的主要性能提升 68
4.1.1 索引去重 68
4.1.2 增量排序 70
4.1.3 库级索引并发重建 71
4.1.4 HASH 聚合可溢出到磁盘 72
4.1.5 索引并行清理 72
4.1.6 PL/pgSQL 提速 74
4.1.7 Windows 连接优化 75

4.2 PostgreSQL 13 的可靠性提高 75
4.2.1 备份可靠性提高 75
4.2.2 流复制可动态配置 77

4.3 PostgreSQL 13 的运维管理优化 77
4.3.1 数据库删除更便捷 77
4.3.2 并行查询关联 PID 78
4.3.3 共享内存可观测 79
4.3.4 基于磁盘的缓存可监控 79
4.3.5 后台操作进度报告引入 80
4.3.6 语句日志采样降噪 80
4.3.7 PSQL 工具跟踪事务运行状态 80
4.3.8 pg_rewind 工具优化 82

4.4 PostgreSQL 13 的开发易用性提升 83
4.4.1 分区表及逻辑复制改进 83
4.4.2 标识列可忽略用户输入 84
4.4.3 存储列可转换为普通列 85
4.4.4 分页排序可并列排名 86
4.4.5 易用的内置函数引入 87
4.4.6 FF1~FF6 时间格式引入 88
4.4.7 Row 表达式使用 88
4.4.8 视图列名纠正 89

4.5 PostgreSQL 13 的系统层变化 89
4.5.1 系统元数据引入 89
4.5.2 配置参数引入 89
4.5.3 对象标识符类型引入 90
4.5.4 内部术语引入 91
4.5.5 备库升主库流程优化 91
4.5.6 INSERT 操作自动清理调优 92
4.5.7 WAL 用量跟踪 93

|　4.5.8　坏块绕过继续恢复..95
|　4.5.9　外部表安全性提高..95
|　4.5.10　附加模块变化..96
4.6　本章小结..97

第 5 章　PostgreSQL 14 新特性...98

5.1　PostgreSQL 14 的主要性能提升..98
|　5.1.1　高并发连接优化..98
|　5.1.2　紧急清理模式引入..99
|　5.1.3　列级压缩可配置..99
|　5.1.4　逻辑复制改进..101
|　5.1.5　嵌套循环改进..102
|　5.1.6　并行特性增强..103
5.2　PostgreSQL 14 的可靠性提高..104
|　5.2.1　数据结构检测..105
|　5.2.2　备节点可作为恢复源..105
|　5.2.3　密码长度限制取消..107
5.3　PostgreSQL 14 的运维管理优化..107
|　5.3.1　查询 ID 引入..108
|　5.3.2　索引表空间在线移动..111
|　5.3.3　触发器在线重建..112
|　5.3.4　控制客户端连接..113
|　5.3.5　后台操作进度报告引入..114
|　5.3.6　可观测性增强..115
5.4　PostgreSQL 14 的开发易用性提升..115
|　5.4.1　多范围类型引入..115
|　5.4.2　存储过程支持 OUT 模式参数..116
|　5.4.3　新形式的 SQL 函数引入...116
|　5.4.4　JSON 操作功能增强..118
|　5.4.5　递归查询改进..119
|　5.4.6　易用的内置函数引入..122
5.5　PostgreSQL 14 的系统层变化..125
|　5.5.1　系统元数据引入..125
|　5.5.2　系统函数变化..126
|　5.5.3　预置角色变化..126
|　5.5.4　配置参数变化..128
|　5.5.5　客户端 C 驱动改进..129
|　5.5.6　客户端认证安全性提高..130

　　　　5.5.7　附加模块变化 131
　5.6　本章小结 133

第6章　PostgreSQL 15 新特性 135

　6.1　PostgreSQL 15 的主要性能提升 135
　　　　6.1.1　统计信息内存化 135
　　　　6.1.2　增量排序算法改进 136
　　　　6.1.3　WAL 恢复预读取 137
　　　　6.1.4　全页写新增压缩算法 138
　　　　6.1.5　备份效率提高 138
　　　　6.1.6　并行特性增强 140
　6.2　PostgreSQL 15 的可靠性提高 141
　　　　6.2.1　统计信息一致性读取 141
　　　　6.2.2　统一非独占备份模式引入 142
　　　　6.2.3　本地化 Collation 相关增强 142
　　　　6.2.4　流复制支持 UNLOGGED 序列 144
　　　　6.2.5　pg_rewind 工具指定外部配置文件 145
　6.3　PostgreSQL 15 的运维管理优化 145
　　　　6.3.1　服务端本地备份引入 145
　　　　6.3.2　JSON 格式日志引入 147
　　　　6.3.3　创建数据库功能增强 150
　　　　6.3.4　COPY 操作对文本格式增强 150
　　　　6.3.5　执行计划显示信息改进 151
　　　　6.3.6　pg_receivewal 工具压缩功能增强 153
　　　　6.3.7　PSQL 工具优化 153
　6.4　PostgreSQL 15 的开发易用性提升 155
　　　　6.4.1　MERGE 语句引入 155
　　　　6.4.2　NULL 值与 UNIQUE 约束更搭 162
　　　　6.4.3　numeric 数据类型改进 163
　　　　6.4.4　正则表达式函数引入 164
　　　　6.4.5　分区表改进 165
　　　　6.4.6　逻辑复制改进 166
　6.5　PostgreSQL 15 的系统层变化 167
　　　　6.5.1　系统元数据引入 167
　　　　6.5.2　系统函数变化 168
　　　　6.5.3　预置角色变化 169
　　　　6.5.4　配置参数变化 170
　　　　6.5.5　GRANT 命令授权变化 173

		6.5.6	递归查询优化	173
		6.5.7	公共模式安全性提高	174
		6.5.8	视图安全性提高	176
		6.5.9	附加模块变化	177
	6.6		本章小结	177

第 7 章　PostgreSQL 16 新特性 179

	7.1		PostgreSQL 16 的主要性能提升	179
		7.1.1	并行特性增强	179
		7.1.2	预排序性能提升	180
		7.1.3	死元组清理性能提升	180
		7.1.4	其他性能提升	180
	7.2		PostgreSQL 16 的运维管理优化	181
		7.2.1	I/O 统计更详细	181
		7.2.2	pg_dump 工具功能增强	182
		7.2.3	PSQL 工具功能增强	183
		7.2.4	VACUUM 命令及 VACUUMDB 工具功能增强	183
		7.2.5	pg_hba.conf 文件配置更高效	184
		7.2.6	普通用户连接优化	184
		7.2.7	HOT 更新监控增强	185
		7.2.8	不活跃索引监控	185
		7.2.9	便捷的参数化语句分析	186
	7.3		PostgreSQL 16 的开发易用性提升	187
		7.3.1	逻辑复制功能完善	187
		7.3.2	SQL 标准 2023 部分支持引入	188
		7.3.3	SQL/JSON 函数功能增强	189
		7.3.4	数据导入默认值重定义	192
		7.3.5	libpq 协议负载均衡功能引入	193
	7.4		PostgreSQL 16 的系统层变化	193
		7.4.1	版本兼容性变化	193
		7.4.2	系统函数变化	195
		7.4.3	预置角色变化	196
		7.4.4	配置参数变化	196
		7.4.5	初始用户权限优化	197
		7.4.6	成员角色权限变化	197
		7.4.7	附加模块变化	200
	7.5		本章小结	200

第 1 章 PostgreSQL 10 新特性

1.1 PostgreSQL 10 的主要性能提升

每个新发布的 PostgreSQL 版本相对之前的版本都有些性能得以提升。这里聚焦 5 个重大性能的提升，除此之外还有些其他新特性。

1.1.1 数据分区

当数据量多到一定程度的时候，会导致性能问题，这个时候可以进行分表处理。不同分表中的数据不重复，从而提高处理效率。PostgreSQL 10 是引入内置分区的第一个版本。用户刚开始使用继承的方式来实现表分区时，PostgreSQL 10 是第一个提供声明式分区（Declarative Partitioning）的版本。当前分区功能支持自动创建合适的子表约束，以及将对父表的修改路由到子表，其详细功能请参考 1.4.1 节声明式分区引入。

1.1.2 并行特性增强

PostgreSQL 9.6 是第一个支持并行查询的版本。PostgreSQL 10 支持更多的并行特性，让用户可以从多核 CPU 中受益。PostgreSQL 10 支持并行 B-Tree 索引扫描、Bitmap 扫描及 Merge Join，并支持并行 PREPARE 语句和 EXECUTE 语句。

下面演示并行 PREPARE 语句和 EXECUTE 语句的过程。

创建测试表并插入大量数据。

```
create table test as select n from generate_series(1,1000000) n;
```

对测试表进行 COUNT 操作，可以看到能使用并行特性。

```
postgres=# explain(costs off) select count(n) from test;
```

```
               QUERY PLAN
---------------------------------------------
Finalize Aggregate
  -> Gather
        Workers Planned: 2
        -> Partial Aggregate
              -> Parallel Seq Scan on test
(5 rows)
```

执行 PREPARE 语句。

```
postgres=# prepare p1 as select count(n) from test;
PREPARE
```

执行 EXECUTE 语句，可以看到能使用并行特性。

```
postgres=# explain(costs off) execute p1;
              QUERY PLAN
---------------------------------------------
Finalize Aggregate
  -> Gather
        Workers Planned: 2
        -> Partial Aggregate
              -> Parallel Seq Scan on test
(5 rows)
```

1.1.3 统计信息扩展

在运行 SQL 语句时，优化器会做出智能判断来加快查询速度。然而，优化器在很大程度上依赖于某个子句或操作用于返回数据的预估值。PostgreSQL 10 之前的版本只显示单列的统计信息，并不知道多列之间的相关性。

示例如下。

```
create table test as
select i, i % 100 as i_100, i % 500 as i_500
from generate_series(1,100000) i;
```

数据显示如下。

```
postgres=# select * from test order by random() limit 10;
   i   | i_100 | i_500
-------+-------+-------
 42490 |    90 |   490
 19886 |    86 |   386
 57385 |    85 |   385
 85404 |     4 |   404
 57517 |    17 |    17
 61908 |     8 |   408
 81299 |    99 |   299
 61389 |    89 |   389
  7730 |    30 |   230
 94512 |    12 |    12
(10 rows)
```

下面分别对 i_100 和 i_500 这两列进行简单查询。

```
postgres=# explain analyze select * from test where i_100 = 12;
                                QUERY PLAN
---------------------------------------------------------------------------
 Seq Scan on test  (cost=0.00..1791.00 rows=930 width=12) (actual time=0.011..15.919
   rows=1000 loops=1)
   Filter: (i_100 = 12)
   Rows Removed by Filter: 99000
 Planning time: 0.038 ms
 Execution time: 16.042 ms
(5 rows)

postgres=# explain analyze select * from test where i_500 = 404;
                                QUERY PLAN
---------------------------------------------------------------------------
 Seq Scan on test  (cost=0.00..1791.00 rows=199 width=12) (actual time=0.071..16.056
   rows=200 loops=1)
   Filter: (i_500 = 404)
   Rows Removed by Filter: 99800
 Planning time: 0.038 ms
 Execution time: 16.089 ms
(5 rows)
```

可以看到，执行计划的两处 rows 中的预估值与实际值比较接近。

下面同时对 i_100 和 i_500 这两列进行关联查询。

```
postgres=# explain analyze select * from test where i_100 = 8 and i_500 = 408;
                                QUERY PLAN
---------------------------------------------------------------------------
 Seq Scan on test  (cost=0.00..2041.00 rows=2 width=12) (actual time=0.067..18.367
   rows=200 loops=1)
   Filter: ((i_100 = 8) AND (i_500 = 408))
   Rows Removed by Filter: 99800
 Planning time: 0.044 ms
 Execution time: 18.413 ms
(5 rows)
```

上面关联查询的 rows 中的预估值是 2，而实际值是 200，差异较大。这是因为优化器简单地将两列的选择率进行相乘，这会得到错误的执行计划，影响 SQL 语句的性能。

在 PostgreSQL 10 中，可以使用 create statistics 命令对多列组合创建统计信息，帮助优化器处理多列依赖关系，让关联列上的执行计划更准确。

```
create statistics test_stats on i_100, i_500 from test;

analyze test;
```

创建统计信息之后，下面查看查询结果。

```
postgres=# explain analyze select * from test where i_100 = 8 and i_500 = 408;
                                QUERY PLAN
---------------------------------------------------------------------------
 Seq Scan on test  (cost=0.00..2041.00 rows=199 width=12) (actual time=0.065..14.444
   rows=200 loops=1)
   Filter: ((i_100 = 8) AND (i_500 = 408))
   Rows Removed by Filter: 99800
```

```
Planning time: 0.092 ms
Execution time: 14.473 ms
(5 rows)
```

可以看到，执行计划的两处 rows 中的预估值与实际值比较接近。

1.1.4 AFTER 触发器性能增强

对于 AFTER 触发器，一次触发涉及的记录可能很多，PostgreSQL 10 对此操作增加了一个过渡表（Transition Table）特性，用于存储触发器涉及的数据。在真正执行触发器函数时，可以直接利用过渡表进行相应的操作，而不必一条一条地读取，这样性能就有了很大的提升。

下面通过示例进行演示。

创建测试表。

```
create table test (id serial primary key, info varchar);

insert into test (info) values
    ('Rat'),
    ('OX'),
    ('Tiger'),
    ('Rabbit'),
    ('Dragon'),
    ('Snake'),
    ('Horse'),
    ('Sheep'),
    ('Monkey'),
    ('Rooster'),
    ('Dog'),
    ('Pig');
```

创建触发器函数。

```
create function test_trigger() returns trigger
as $$
declare
   rec record;
begin
   for rec in select * from v_old_table LOOP
      raise notice 'OLD: %', rec;
   end loop;
   for rec in select * from v_new_table LOOP
      raise notice 'NEW: %', rec;
   end loop;
   return NEW;
end;
$$ language plpgsql;
```

上面的函数在触发器调用期间，会显示 v_old_table 表和 v_new_table 表中的所有记录，而实际上这里只有测试表。

```
postgres=# \d
            List of relations
 Schema |   Name   |  Type  | Owner
```

```
--------+--------------+------------+----------
 public | test         | table      | postgres
 public | test_id_seq  | sequence   | postgres
(2 rows)
```

创建触发器。

```
create trigger tg_test
   after update on test
   referencing new table as v_new_table
       old table as v_old_table
   for each statement
   execute procedure test_trigger();
```

测试 UPDATE 操作更新单条记录。

```
postgres=# update test set info = 'test' where id = 1;
NOTICE:  OLD: (1,Rat)
NOTICE:  NEW: (1,test)
UPDATE 1
```

测试 UPDATE 操作更新多条记录。

```
postgres=# update test set info = info || '_new' where id > 8;
NOTICE:  OLD: (9,Monkey)
NOTICE:  OLD: (10,Rooster)
NOTICE:  OLD: (11,Dog)
NOTICE:  OLD: (12,Pig)
NOTICE:  NEW: (9,Monkey_new)
NOTICE:  NEW: (10,Rooster_new)
NOTICE:  NEW: (11,Dog_new)
NOTICE:  NEW: (12,Pig_new)
UPDATE 4
```

可以使用 v_old_table 表和 v_new_table 表对每条语句触发时涉及的记录进行处理。

在生产环境的一个事务里频繁对单行数据进行 UPDATE 操作会引发 VACUUM 操作问题,此时基于触发器统计更新数据会变得越来越慢。对比,可以使用触发器的过渡表特性,使用一条语句进行批量处理。

1.1.5 Aggregate 操作支持下推

PostgreSQL 10 对外部表的重大改进有 Aggregate 操作支持下推,FDW 不需要先去远程服务器上获取所有数据行再进行本地合并。下面分别在 PostgreSQL 9.6 和 PostgreSQL 10 中进行测试。

在 PostgreSQL 9.6 中进行测试。

```
postgres=# create table local_tab as select i as id, cast(random() * 10 as int4) as
   group_id from generate_series(1,100000) i;
SELECT 100000

postgres=# create extension postgres_fdw;
CREATE EXTENSION
```

```
postgres=# create server local_pg_server foreign data wrapper postgres_fdw options
    (dbname 'postgres');
CREATE SERVER

postgres=# create user mapping for current_user server local_pg_server options(user
    'postgres');
CREATE USER MAPPING

postgres=# create foreign table remote_tab (id int4, group_id int4) server
    local_pg_server options (table_name 'local_tab');
CREATE FOREIGN TABLE
```

检查数据。

```
postgres=# select group_id, count(*) from local_tab group by group_id order by
    group_id;
 group_id | count
----------+-------
        0 |  5074
        1 |  9923
        2 |  9940
        3 | 10050
        4 |  9978
        5 |  9878
        6 |  9975
        7 | 10082
        8 | 10131
        9 |  9995
       10 |  4974
(11 rows)

postgres=# select group_id, count(*) from remote_tab group by group_id order by
    group_id;
 group_id | count
----------+-------
        0 |  5074
        1 |  9923
        2 |  9940
        3 | 10050
        4 |  9978
        5 |  9878
        6 |  9975
        7 | 10082
        8 | 10131
        9 |  9995
       10 |  4974
(11 rows)
```

本地表与外部表的数据是一致的。查看 PostgreSQL 9.6 中语句的执行计划。

```
postgres=# explain (analyze,verbose) select group_id, count(*) from remote_tab
    group by group_id order by group_id;
                          QUERY PLAN
```

```
GroupAggregate  (cost=100.00..222.22 rows=200 width=12) (actual
  time=129.733..275.104 rows=11 loops=1)
  Output: group_id, count(*)
  Group Key: remote_tab.group_id
  -> Foreign Scan on public.remote_tab  (cost=100.00..205.60 rows=2925 width=4)
  (actual time=121.830..263.083 rows=100000 loops=1)
        Output: id, group_id
        Remote SQL: SELECT group_id FROM public.local_tab ORDER BY group_id ASC
  NULLS LAST
Planning time: 0.051 ms
Execution time: 275.740 ms
(8 rows)
```

观察相同的语句在 PostgreSQL 10 中的执行计划。

```
postgres=# explain (analyze,verbose,costs off) select group_id, count(*) from
  remote_tab group by group_id order by group_id;
                        QUERY PLAN
Sort  (cost=167.52..168.02 rows=200 width=12) (actual time=27.871..27.872 rows=11
  loops=1)
  Output: group_id, (count(*))
  Sort Key: remote_tab.group_id
  Sort Method: quicksort  Memory: 25kB
  -> Foreign Scan  (cost=114.62..159.88 rows=200 width=12) (actual
  time=27.860..27.862 rows=11 loops=1)
        Output: group_id, (count(*))
        Relations: Aggregate on (public.remote_tab)
        Remote SQL: SELECT group_id, count(*) FROM public.local_tab GROUP BY 1
Planning time: 0.075 ms
Execution time: 28.080 ms
(10 rows)
```

可以发现，第二个执行计划比第一个执行计划在 Foreign Scan 中多了 Relations：Aggregate on（对外部表执行合并操作），另外，对于 actual…rows，在 PostgreSQL 9.6 中值为 100000，在 PostgreSQL 10 中值为 11。很明显，在 PostgreSQL 10 中通过网络传输的数据量更少，从最后的执行时间中也能很好地反映出这个差异。

1.2 PostgreSQL 10 的可靠性提高

PostgreSQL 10 的可靠性提高主要体现在仲裁提交引入、HASH（哈希）索引支持记录 WAL（Write-Ahead Log）、事务提交状态检测。

1.2.1 仲裁提交引入

在 PostgreSQL 10 之前的版本中只能有一台服务器充当同步节点，现在引入了仲裁提交。这个设计实际上非常简单。假设希望 7 台服务器中的 5 台来确认事务提交，这就是仲裁提交要表达的想法，它允许开发人员或管理员以更优雅的方式来定义事务提交。

仲裁提交对参数 synchronous_standby_names 的语法进行了扩展，支持使用 FIRST 和 ANY 两种模式指定同步复制节点。下面是两个配置示例。

```
synchronous_standby_names = FIRST 1 (s1, s2)
synchronous_standby_names = ANY 2 (s1, s2, s3)
```

1.2.2　HASH 索引支持记录 WAL

在很长的一段时间中，使用 HASH 索引是不太安全的，这是因为 HASH 索引不记录 WAL，不能发送到备库，在 PostgreSQL 10 之前的版本中如果对表创建 HASH 索引，那么会提示警告信息，示例如下。

在 PostgreSQL 9.6 中对表创建 HASH 索引，提示警告信息。

```
postgres=# create index idx_hash on tab_test using hash (i) ;
WARNING:  hash indexes are not WAL-logged and their use is discouraged
CREATE INDEX
```

在 PostgreSQL 10 中对表创建 HASH 索引，不提示警告信息。

```
postgres=> CREATE INDEX idx1_hash1 ON hash1 USING hash (c1) ;
CREATE INDEX
```

另外，使用 PostgreSQL 9.6 在备库中进行查询时，如果查询使用索引，那么可能会读不到文件，PostgreSQL 10 及之后的版本不会存在这个问题了，HASH 索引现在可以安全地使用。

1.2.3　事务提交状态检测

在有些情况下，当客户端提交事务后，如果客户端在接收数据库返回的提交状态前，数据库崩溃或客户端断开连接，那么客户端就不知道事务是否提交成功。PostgreSQL10 引入了一个新功能，可以查看以往的事务状态。

事务状态通过 txid_status(bigint) 函数获取，该函数的参数为要获取的事务状态的事务号。客户端只有保留此事务号，才可以获取该事务状态。

```
postgres=# begin;
BEGIN

postgres=# select txid_current();
 txid_current
--------------
          575
(1 row)


postgres=# select txid_status(575);
 txid_status
-------------
 in progress
(1 row)
```

```
postgres=# commit;
COMMIT

postgres=# select txid_status(575);
 txid_status
-------------
 committed
(1 row)

postgres=# select txid_status(800);
ERROR:  transaction ID 800 is in the future
```

1.3　PostgreSQL 10 的运维管理优化

PostgreSQL 10 的运维管理优化主要体现在 WAL 文件大小扩容、WAL 支持在线压缩、ICU（Internation Component for Unicode）标准库引入、活动会话视图增强、HBA 文件新增系统视图、promote 子命令增加等待模式、基础备份增强、临时复制槽引入、行级安全策略优化、PSQL 工具允许条件式交互。

1.3.1　WAL 文件大小扩容

PostgreSQL 源码在编译配置 configure 命令时，可以使用--with-wal-segsize 选项设置 WAL 文件的大小。

PostgreSQL 9.6 可以设置 WAL 文件的大小为 1MB、2MB、4MB、8MB、16MB、32MB、64MB，最大为 64MB。

```
$ /opt/postgresql-9.6.24/configure --with-wal-segsize=128
checking build system type... x86_64-pc-linux-gnu
checking host system type... x86_64-pc-linux-gnu
checking which template to use... linux
checking whether to build with 64-bit integer date/time support... yes
checking whether NLS is wanted... no
checking for default port number... 5432
checking for block size... 8kB
checking for segment size... 1GB
checking for WAL block size... 8kB
checking for WAL segment size... configure: error: Invalid WAL segment size.
   Allowed values are 1,2,4,8,16,32,64.
```

PostgreSQL 10 可以设置 WAL 文件的大小为 1MB、2MB、4MB、8MB、16MB、32MB、64MB、128MB、256MB、512MB、1024MB，最大为 1024MB。

```
$ /opt/postgresql-10.18/configure --with-wal-segsize=2048
checking build system type... x86_64-pc-linux-gnu
checking host system type... x86_64-pc-linux-gnu
checking which template to use... linux
```

```
checking whether NLS is wanted... no
checking for default port number... 5432
checking for block size... 8kB
checking for segment size... 1GB
checking for WAL block size... 8kB
checking for WAL segment size... configure: error: Invalid WAL segment size.
  Allowed values are 1,2,4,8,16,32,64,128,256,512,1024.
```

1.3.2　WAL 支持在线压缩

在 PostgreSQL 10 中备份 WAL 时，推荐在归档服务器上使用 pg_receivewal 工具备份 WAL。使用 pg_receivewal 工具备份 WAL 比使用归档命令备份 WAL 更加安全，不需要等待 WAL 文件写满，只需要基于流复制协议流式传输 WAL，并且当重启服务器时，不会发生数据丢失或数据损坏的情况。PostgreSQL 10 中的 pg_receivewal 工具新增了 -Z/--compress 选项，支持对 WAL 进行在线压缩，参数值范围为 0~9，0 表示无压缩，9 表示最大压缩，示例如下：

```
$ pg_receivewal --host=x.x.x.x --compress=9 --directory=archive &
```

1.3.3　ICU 标准库引入

当创建 PostgreSQL 10 时，管理员可以选择设置编码，编码决定着存储的字符的排列顺序。PostgreSQL 10 依赖操作系统实现这种排列顺序。如果操作系统中字符的排列顺序在某些时间内发生变化（可能是因为一个 BUG 或其他原因），那么 PostgreSQL 的索引会产生问题。

引入 ICU 标准库可以解决这个问题。ICU 标准库可以提供比操作系统更牢固的保障，适合长期存储数据。PostgreSQL10 开始支持 ICU 标准库，使用前需要安装 libicu 依赖包和 libicu-devel 依赖包，同时在编译数据库时需要使用 --with-icu 选项。

```
$ ./configure --with-icu --prefix=/opt/pg10
```

1.3.4　活动会话视图增强

PostgreSQL 10 的活动会话视图（pg_stat_activity 视图）新增了 backend_type 字段，增加了一些系统服务进程信息，而 PostgreSQL 9.6 中的该视图只包括用户后台服务进程。

```
postgres=# \x
Expanded display is on.

postgres=# SELECT pid, wait_event_type, wait_event, backend_type
 FROM pg_stat_activity ;
-[ RECORD 1 ]---+--------------------
pid             | 12159
wait_event_type | Activity
wait_event      | AutoVacuumMain
backend_type    | autovacuum launcher
-[ RECORD 2 ]---+--------------------
```

```
pid             | 12161
wait_event_type | Activity
wait_event      | LogicalLauncherMain
backend_type    | background worker
-[ RECORD 3 ]---+--------------------
pid             | 12628
wait_event_type |
wait_event      |
backend_type    | client backend
-[ RECORD 4 ]---+--------------------
pid             | 12156
wait_event_type | Activity
wait_event      | BgWriterMain
backend_type    | background writer
-[ RECORD 5 ]---+--------------------
pid             | 12155
wait_event_type | Activity
wait_event      | CheckpointerMain
backend_type    | checkpointer
-[ RECORD 6 ]---+--------------------
pid             | 12157
wait_event_type | Activity
wait_event      | WalWriterMain
backend_type    | walwriter
```

1.3.5 HBA 文件新增系统视图

pg_hba.conf 文件是 PostgreSQL 10 控制客户端访问的核心配置文件，是数据库层的防火墙，可以通过设置白名单或黑名单进行网络连接加固。PostgreSQL 10 在 SQL 接口增加了 pg_hba_file_rules 视图。下面是 pg_hba_file_rules 视图的输出示例。

```
postgres=# SELECT line_number, type, database, user_name, address, auth_method FROM
    pg_hba_file_rules;
 line_number |  type  |   database    | user_name |  address  | auth_method
-------------+--------+---------------+-----------+-----------+-------------
          84 | local  | {all}         | {all}     |           | trust
          86 | host   | {all}         | {all}     | 127.0.0.1 | trust
          88 | host   | {all}         | {all}     | ::1       | trust
          91 | local  | {replication} | {all}     |           | trust
          92 | host   | {replication} | {all}     | 127.0.0.1 | trust
          93 | host   | {replication} | {all}     | ::1       | trust
(6 rows)
```

pg_hba_file_rules 视图的 error 字段可以追踪 pg_hba.conf 文件中每行配置的问题。例如，pg_hba.conf 文件的下面这行配置忘记配置掩码位数。

```
host    all             all             192.168.137.250             trust
```

当发送 reload 信号之后，进入数据库查询，会看到明显的提示信息。

```
postgres=# \x
Expanded display is on.
```

```
-[ RECORD 1 ]--------------------------------------------------
line_number | 87
type        |
database    |
user_name   |
address     |
auth_method |
error       | invalid IP mask "trust": Name or service not known
```

1.3.6 promote 子命令增加等待模式

pg_ctl 命令用于管理 PostgreSQL 服务，它的一些子命令可以选择是否等待操作完成后再退出 pg_ctl 命令。PostgreSQL 9.6 默认为 shutdown 子命令时会等待，而为 start 子命令和 restart 子命令时则不会等待，可以通过-w 选项进行切换。在 PostgreSQL 10 中，promote 子命令也拥有这个权益，并且所有子命令都默认为会等待。

使用 PostgreSQL 9.6 查看 pg_ctl 命令的关键字 wait，只有默认为 shutdown 子命令时会等待，如下所示。

```
$ pg_ctl --help |grep wait
  -t, --timeout=SECS     seconds to wait when using -w option
  -w                     wait until operation completes
  -W                     do not wait until operation completes
(The default is to wait for shutdown, but not for start or restart.)
```

使用 PostgreSQL 10 查看 pg_ctl 命令的关键字 wait，默认为等待，如下所示。

```
$ pg_ctl --help |grep wait
  -t, --timeout=SECS     seconds to wait when using -w option
  -w, --wait             wait until operation completes (default)
  -W, --no-wait          do not wait until operation completes
```

1.3.7 基础备份增强

pg_basebackup 工具在 PostgreSQL 9.6 中备份 WAL 的默认方式是 fetch 方式，这种方式不如 stream 方式安全。在 PostgreSQL 10 中备份 WAL 的默认方式是 stream 方式，允许同时使用 stream 方式备份 WAL 与输出 tar 包格式。

在 PostgreSQL 9.6 中，不允许同时使用 stream 方式备份 WAL 与输出 tar 包格式。

```
$ pg_basebackup --pgdata=data_backup --format=t --xlog-method=stream
pg_basebackup: WAL streaming can only be used in plain mode
Try "pg_basebackup --help" for more information.
```

在 PostgreSQL 10 中，允许同时使用 stream 方式备份 WAL 与输出 tar 包格式。

```
$ pg_basebackup --pgdata=data_backup --format=t --wal-method=stream

$ ls -lh data_backup/
total 116M
-rw-r--r-- 1 postgres postgres 100M Jun 20 19:07 base.tar
-rw------- 1 postgres postgres  17M Jun 20 19:07 pg_wal.tar
```

1.3.8 临时复制槽引入

PostgreSQL 10 引入了临时复制槽，功能和正常复制槽一样，只是临时复制槽在会话结束后被自动删除。使用临时复制槽可以降低复制过程中存在的风险。在数据库集簇中，不允许同名的复制槽存在，如果一个复制槽在使用后，没有被及时删除，而后来要创建的复制槽正好与之同名，那么复制会失败。

临时复制槽可以用于使用 pg_basebackup 工具进行基础备份及搭建流复制环境。

```
$ pg_basebackup --pgdata=data_backup --wal-method=stream
```

在使用 stream 方式进行基础备份时，会自动创建临时复制槽，可以设置参数 log_replication_commands 的值为 on，并可以通过数据库日志观察到以 pg_basebackup 为前缀命名的临时复制槽的创建过程。

```
YYYY-MM-DD HH:MM:SS.SSS CST,"postgres","",6197,"[local]",62b13348.1835,2,
"idle",YYYY-MM-DD HH:MM:SS CST,4/0,0,LOG,00000,"received replication command:
  CREATE_REPLICATION_SLOT ""pg_basebackup_6197"" TEMPORARY PHYSICAL
  RESERVE_WAL",,,,,,"CREATE_REPLICATION_SLOT ""pg_basebackup_6197"" TEMPORARY
  PHYSICAL RESERVE_WAL",,,"pg_basebackup"
```

1.3.9 行级安全策略优化

在 PostgreSQL 9.6 及 PostgreSQL 9.6 之前的版本中，当对一个表创建多个策略时，创建的策略始终是宽容性（PERMISSIVE）策略，也就是说，当多个策略都适用时通过 OR 组合在一起。在 PostgreSQL 10 中，当对一个表创建多个策略时新增了限制性（RESTRICTIVE）策略，也就是说，当多个策略都适用时通过 AND 组合在一起。

下面测试宽容性策略和限制性策略组合的影响，通过对 tab_pcy 表创建宽容性策略 pcy1，以及限制性策略 pcy2 和 pcy3，来验证策略组合。

创建 tab_pcy 表，以及 pcy1 策略和 pcy2 策略。

```
postgres=> create table tab_pcy (c1 int, c2 varchar, uname varchar) ;
CREATE TABLE

postgres=> alter table tab_pcy enable row level security ;
ALTER TABLE

postgres=> create policy pcy1 on tab_pcy for all using (uname = current_user);
CREATE POLICY

postgres=> create policy pcy2 on tab_pcy as restrictive for all using (c2 =
   'data') ;
CREATE POLICY

postgres=> \d tab_pcy
             Table "public.tab_pcy"
 Column |       Type        | Collation | Nullable | Default
--------+-------------------+-----------+----------+---------
 c1     | integer           |           |          |
```

```
    c2      | character varying |           |          |
    uname   | character varying |           |          |
    POLICY "pcy1"
      USING (((uname)::name = CURRENT_USER))
    POLICY "pcy2" AS RESTRICTIVE
      USING (((c2)::text = 'data'::text))
```

可以看出，tab_pcy 表的 pcy2 策略是限制性策略。

查看执行计划。

```
postgres=> explain(costs off) select * from tab_pcy ;
                       QUERY PLAN
------------------------------------------------
 Seq Scan on tab_pcy
   Filter: (((c2)::text = 'data'::text) AND ((uname)::name = CURRENT_USER))
(2 rows)
```

可以看出，pcy1 策略和 pcy2 策略通过 AND 组合在一起。

删除 pcy1 策略，并创建 pcy3 策略。

```
postgres=> create policy pcy3 on tab_pcy as restrictive for all using (c1 > 100) ;
CREATE POLICY

postgres=> \d tab_pcy
              Table "public.tab_pcy"
 Column  |       Type        | Collation | Nullable | Default
---------+-------------------+-----------+----------+---------
 c1      | integer           |           |          |
 c2      | character varying |           |          |
 uname   | character varying |           |          |
Policies:
    POLICY "pcy2" AS RESTRICTIVE
      USING (((c2)::text = 'data'::text))
    POLICY "pcy3" AS RESTRICTIVE
      USING ((c1 > 100))
```

查看执行计划。

```
postgres=> explain(costs off) select * from tab_pcy ;
      QUERY PLAN
-------------------------
 Result
   One-Time Filter: false
(2 rows)
```

如果所有策略都是限制性策略，那么执行计划将无显示信息。

宽容性策略和限制性策略，即行级安全策略组合测试的结果如表 1-1 所示。

表 1-1 行级安全策略组合测试的结果

策略名	策略组合一	策略组合二	策略组合三	策略组合四
pcy1	PERMISSIVE	PERMISSIVE	RESTRICTIVE	RESTRICTIVE
pcy2	PERMISSIVE	RESTRICTIVE	PERMISSIVE	RESTRICTIVE
结果	OR	AND	AND	NULL

1.3.10　PSQL 工具允许条件式交互

PostgreSQL 10 中的 PSQL 工具比较重大的变化是引入了条件式交互语法\if...\elif...\else...\endif...，可以在 PSQL 工具中与数据库进行条件式交互。

在 PostgreSQL 10 之前的版本中可以通过下面的方式进行分区判断和处理。

```
postgres => create temporary table my_tab as select 1 as x;
SELECT 1

postgres=> select case
postgres->             when exists(select null from my_tab)
postgres->             then '\echo not empty \q'
postgres->             else 'drop table my_table;'
postgres->          end as cmd
postgres-> \gset
postgres=> :cmd
not empty
```

很明显，使用上面的方式表述不是特别清晰，很难实现多个复杂的操作，并且很难实现嵌套操作。

在 PostgreSQL 10 中，可以通过条件式交互进行非常灵活的扩展操作，同时也可以进行嵌套操作，下面使用上述示例中的 case 语句在 PostgreSQL 10 中进行条件式交互。

```
postgres=> select exists (select null from pg_class
postgres(>              where relname = 'my_tab') as my_table_exists
postgres-> \gset
postgres=> \if :my_table_exists
postgres@> drop table my_table;
query ignored; use \endif or Ctrl-C to exit current \if block
postgres-> \endif
```

1.4　PostgreSQL 10 的开发易用性提升

PostgreSQL 10 的开发易用性提升主要体现在声明式分区引入、表级数据发布与订阅、标识列引入、全文检索支持 JSON 与 JSONB 数据类型、xmltable 函数引入。

1.4.1　声明式分区引入

分区是把逻辑上的一个大表分裂成多个更小的物理分片，以获得性能的提升。PostgreSQL 10 引入了声明式分区，在此之前，用户只能使用表继承的方式。声明式分区支持以下两种方式的分区。

1．范围分区

范围分区（Range Partitioning）把一个或一组字段作为分区键，将表划分成多个范围，分配给不同分区的值的范围之间没有重叠。

例如，已知有一个销售表，下面按销售日期对销售表进行分区。

```
create table sales(
    id serial,
    sales_count int,
    sales_date date not null
) partition by range(sales_date);
```

可以看出，销售表按销售日期范围进行了分区。

创建具体销售日期范围的分区。

```
create table sales_2020_01 partition of sales for values from ('2020-01-01') to ('2020-02-01');
create table sales_2020_02 partition of sales for values from ('2020-02-01') to ('2020-03-01');
create table sales_2020_03 partition of sales for values from ('2020-03-01') to ('2020-04-01');
create table sales_2020_04 partition of sales for values from ('2020-04-01') to ('2020-05-01');
create table sales_2020_05 partition of sales for values from ('2020-05-01') to ('2020-06-01');
create table sales_2020_06 partition of sales for values from ('2020-06-01') to ('2020-07-01');
create table sales_2020_07 partition of sales for values from ('2020-07-01') to ('2020-08-01');
create table sales_2020_08 partition of sales for values from ('2020-08-01') to ('2020-09-01');
create table sales_2020_09 partition of sales for values from ('2020-09-01') to ('2020-10-01');
create table sales_2020_10 partition of sales for values from ('2020-10-01') to ('2020-11-01');
create table sales_2020_11 partition of sales for values from ('2020-11-01') to ('2020-12-01');
create table sales_2020_12 partition of sales for values from ('2020-12-01') to ('2021-01-01');
```

可以看出，按月份对 2020 年创建了 12 个分区。

2．列表分区

列表分区（List Partitioning）根据一个分区字段（只能有一个分区字段或表达式），列举该字段的某个值或某几个值归属于某个分区，即分别对每个分区罗列出属于该分区的字段值列表。

例如，对人员信息表中的国籍设置列表值。

```
create table person(
    personname   varchar(100),
    country      varchar(100)
) partition by list(country);
```

下面创建不同国家组成的区域人员的分区。

```
create table asia partition of person for values in('CHINA','INDIA');
create table africa partition of person for values in('ANGOLA','BENIN');
create table northamerica partition of person for values in('US','MEXICO');
```

```
create table australia partition of person for values in('AUSTRALIA','FIJI');
```

注意，表分区在 PostgreSQL 10 中有如下限制。
- 不允许创建全局索引、主键约束、唯一约束、外键约束和排他约束。
- 不支持默认分区。
- 普通表不允许外键引用分区表中的字段。
- 不允许全局使用 ON CONFLICT 子句，这是因为不支持全局唯一约束。
- 数据更新不能跨分区移动。
- 不允许全局使用行级触发器，分区上的行级触发器不支持过渡表。
- 普通表不能继承分区父表和子表。

1.4.2 表级数据发布与订阅

PostgreSQL 10 开始支持逻辑复制，可以在多个实例之间进行部分表中数据的复制。在传统的物理流复制场景下，从节点只能提供只读服务，但表级逻辑复制支持读写服务。

逻辑复制使用订阅与发布语法，发布数据使用 CREATE PUBLICATION 命令。

```
postgres=# \h CREATE PUBLICATION
Command: CREATE PUBLICATION
Description: define a new publication
Syntax:
CREATE PUBLICATION name
 [ FOR TABLE [ ONLY ] table_name [ * ] [, ...]
 | FOR ALL TABLES ]
 [ WITH ( publication_parameter [= value] [, ... ] ) ]
```

下面是几个发布示例：第一个示例是发布两个表的 INSERT 操作、UPDATE 操作、DELETE 操作；第二个示例是发布所有表的 INSERT 操作、UPDATE 操作、DELETE 操作；第三个示例是发布单个表的 INSERT 操作。

```
create publication pub_two for table tab1, tab2;

create publication pub_all for all tables;

create publication pub_tab1_insert for table tab1 with(publish = 'insert');
```

一旦数据集被发布，远程服务器就可以进行订阅并接收发布的数据集。

```
postgres=# \h CREATE SUBSCRIPTION
Command: CREATE SUBSCRIPTION
Description: define a new subscription
Syntax:
CREATE SUBSCRIPTION subscription_name
 CONNECTION 'conninfo'
 PUBLICATION publication_name [, ...]
 [ WITH ( subscription_parameter [= value] [, ... ] ) ]
```

下面是几个订阅示例：第一个示例是创建订阅并立即开始复制；第二个示例是创建订阅并稍后开始复制；第三个示例是在同一个实例下使用自定义复制槽创建订阅。

```
create subscription mysub1 CONNECTION 'host=x.x.x.x port=5432 user=foo
   dbname=foodb' publication pub_all;
```

```
create subscription mysub2 CONNECTION 'host=x.x.x.x port=5432 user=foo
   dbname=foodb' publication pub_tab1_insert WITH (enabled = false);

create subscription mysub CONNECTION 'host=127.0.0.1 port=5432 user=foo
   dbname=foodb' publication pub_tab1_insert
WITH (create_slot='false',slot_name='myslot');
```

逻辑复制设计的精妙之处在于在服务器上既可以发布数据集又可以同时订阅数据集，这里不存在主从角色一说。使用逻辑复制时，分发数据更加便利。

注意，在 PostgreSQL 10 中使用逻辑复制有如下限制。

- 支持 INSERT 操作、UPDATE 操作、DELETE 操作的数据复制，对表进行 CREATE 操作、ALTER 操作时，不会复制到远端。
- 逻辑复制只针对表对象，不支持表的索引、触发器复制到远端。
- 普通表不允许外键引用分区表中的字段。
- 在复制过程中，如果产生冲突，那么会暂停复制。
- 视图、物化视图、分区父表及外部表不能进行复制。

1.4.3 标识列引入

标识列是 SQL 标准 serial 伪类型的一种变体。其功能和序列类似，并且修复了序列存在的以下问题。

- 在使用 CREATE TABLE / LIKE 命令复制表时引用相同的序列。
- 不能使用 ALTER TABLE 命令增加或删除序列属性。
- 在删除默认值 default 时不会关联删除序列。
- 序列需要进行额外赋权管理。

下面通过示例进行说明。

创建一个基于序列主键的 test_serial 表。

```
create table test_serial (
   id serial primary key,
   info text
);

postgres=# \d test_serial
                  Table "public.test_serial"
 Column |  Type   | Collation | Nullable |              Default
--------+---------+-----------+----------+------------------------------------
 id     | integer |           | not null | nextval('test_serial_id_seq'::regclass)
 info   | text    |           |          |
Indexes:
"test_serial_pkey" PRIMARY KEY, btree (id)
```

创建一个基于标识列的 test_identity 表。

```
create table test_identity (
   id int generated by default as identity primary key,
```

```
    info text
);

postgres=# \d test_identity
                Table "public.test_identity"
 Column |  Type   | Collation | Nullable |            Default
--------+---------+-----------+----------+---------------------------------
 id     | integer |           | not null | generated by default as identity
 info   | text    |           |          |
Indexes:
    "test_identity_pkey" PRIMARY KEY, btree (id)
```

对比两个表的结构可以看出，默认值有一点差异，而 INSERT 操作是一样的。

```
postgres=# insert into test_serial(info) values('a'),('b'),('c') returning *;
 id | info
----+------
  1 | a
  2 | b
  3 | c
(3 rows)

INSERT 0 3

postgres=# insert into test_identity(info) values('a'),('b'),('c') returning *;
 id | info
----+------
  1 | a
  2 | b
  3 | c
(3 rows)

INSERT 0 3
```

除了 INSERT 操作，UPDATE 操作和 DELETE 操作的使用也没什么区别。

观察表中创建的序列。

```
postgres=# \ds
                List of relations
 Schema |         Name         |   Type   |  Owner
--------+----------------------+----------+----------
 public | test_identity_id_seq | sequence | postgres
 public | test_serial_id_seq   | sequence | postgres
(2 rows)
```

下面测试删除两个表使用的序列。

测试删除 test_serial 表使用的序列。

```
postgres=# drop sequence test_serial_id_seq cascade ;
NOTICE:  drop cascades to default value for column id of table test_serial
DROP SEQUENCE

postgres=# \d test_serial
           Table "public.test_serial"
 Column |  Type   | Collation | Nullable | Default
```

```
 --------+---------+-----------+----------+---------
 id      | integer |           | not null |
 info    | text    |           |          |
Indexes:
"test_serial_pkey" PRIMARY KEY, btree (id)
```

测试删除 test_identity 表使用的序列。

```
postgres=# drop sequence test_identity_id_seq cascade ;
ERROR:  cannot drop sequence test_identity_id_seq because table test_identity
   column id requires it
HINT:  You can drop table test_identity column id instead.
```

即使使用 cascade 选项也不能删除使用的序列，必须先删除标识列。

在创建 test_identity 表时，使用 generated by default as identity 可以对标识列显式赋值，即覆盖系统赋值，手动插入。而使用 generated always as identity 则不能对标识列显式赋值，否则会报错。

在创建 test_identity 表时，使用 generated by default as identity 可以对标识列显式赋值。

```
postgres=# insert into test_identity values(4,'d');
INSERT 0 1
```

在创建 test_identity2 表时，使用 generated always as identity 不能对标识列显式赋值，否则会报错。

```
create table test_identity2 (
    id int generated always as identity primary key,
    info text
);

postgres=# insert into test_identity2 values(99,'abc');
ERROR:  cannot insert into column "id"
DETAIL:  Column "id" is an identity column defined as GENERATED ALWAYS.
HINT:  Use OVERRIDING SYSTEM VALUE to override.
```

在进行 INSERT 操作时，使用 overriding system value 可以对标识列强制赋值。

```
postgres=# insert into test_identity2(id, info) overriding system value
 values (99, 'abc') returning *;
 id | info
----+------
 99 | abc
(1 row)

INSERT 0 1
```

在进行 UPDATE 操作时，对标识列强制赋值与 INSERT 操作类似，也会报错。

```
postgres=# update test_identity2 set id=199 where id=99;
ERROR:  column "id" can only be updated to DEFAULT
DETAIL:  Column "id" is an identity column defined as GENERATED ALWAYS.
```

标识列修复了序列存在的一些问题，并且支持使用 generated by default as identity 和 generated always as identity 两种形式来控制用户对标识列显式赋值还是系统自动赋值，同时系统自动赋值后也可以再次使用 overriding system value 来强制赋值，否则在进行 INSERT 操作及 UPDATE 操作时显式赋值会报错。

1.4.4 全文检索支持 JSON 与 JSONB 数据类型

从 PostgreSQL 10 开始，全文检索支持 JSON 与 JSONB 数据类型，新增了 to_tsvector 函数和 ts_headline 函数。

使用 to_tsvector 函数对 JSON 与 JSONB 数据类型构建索引。

```
postgres=# create table test_json(id int,a json,b jsonb);
CREATE TABLE

postgres=# create index idx1 on test_json using gin((to_tsvector('english',a)));
CREATE INDEX

postgres=# create index idx2 on test_json using gin((to_tsvector('english',b)));
CREATE INDEX
```

使用 ts_headline 函数对搜索结果加亮显示。

```
postgres=# select jsonb_pretty(jsonlog::jsonb) from test_jsonlog;
                         jsonb_pretty
-----------------------------------------------------------------
 {                                                              +
     "pid": 14907,                                              +
     "hint": "Future log output will go to log destination \"jsonlog\".",+
     "txid": 0,                                                 +
     "message": "ending log output to stderr",                  +
     "line_num": 1,                                             +
     "query_id": 0,                                             +
     "timestamp": "2022-07-01 07:45:24.803 CST",                +
     "session_id": "62be3594.3a3b",                             +
     "backend_type": "postmaster",                              +
     "session_start": "2022-07-01 07:45:24 CST",                +
     "error_severity": "LOG"                                    +
 }
(1 row)
```

对关键字 stderr 使用 ts_headline 函数加亮显示。

```
postgres=# select jsonb_pretty(ts_headline(jsonlog::jsonb, 'stderr'::tsquery)) from
  test_jsonlog;
                         jsonb_pretty
-----------------------------------------------------------------
 {                                                              +
     "pid": 14907,                                              +
     "hint": "Future log output will go to log destination \"jsonlog\".",+
     "txid": 0,                                                 +
     "message": "ending log output to <b>stderr</b>",           +
     "line_num": 1,                                             +
     "query_id": 0,                                             +
     "timestamp": "2022-07-01 07:45:24.803 CST",                +
     "session_id": "62be3594.3a3b",                             +
     "backend_type": "postmaster",                              +
     "session_start": "2022-07-01 07:45:24 CST",                +
     "error_severity": "LOG"                                    +
```

```
}
(1 row)
```

注意，关键字 stderr 前后包裹了一对标签。

1.4.5 xmltable 函数引入

xmltable 函数是 SQL/XML 标准定义构造 XML 格式数据的函数，允许将 XML 格式数据当成表一样与其他表进行关联查询，并且 xmltable 函数对 XML 格式数据的处理性能比客户端的处理性能高。

下面构造一个简单的 XML 格式表，作为数据源。

```
create table test_people as select
xml $$
<people>
    <person>
        <first_name>Pavel</first_name>
        <last_name>Feng</last_name>
        <nick>john</nick>
    </person>
    <person>
        <first_name>Jerome</first_name>
        <last_name>Peng</last_name>
    </person>
</people>
$$ as xml_data;
```

将 XML 格式数据当成表进行查询。

```
select decoded.*
from
    test_people,
    xmltable(
        '//people/person'
        passing xml_data
        columns
            first_name text,
            last_name text,
            nick text
    ) as decoded;
```

查询结果如下。

```
 first_name | last_name | nick
------------+-----------+------
 Pavel      | Feng      | john
 Jerome     | Peng      |
(2 rows)
```

对 XML 格式数据中的节点进行重命名。

```
select decoded.*
from
    test_people,
    xmltable(
```

```
            '//people/person'
            passing xml_data
            columns
                first_name text,
                last_name text,
                nick_name text path 'nick'
    ) as decoded;
```

查询结果如下。在返回表字段时,nick 映射为 nick_name。

```
 first_name | last_name | nick_name
------------+-----------+-----------
 Pavel      | Feng      | john
 Jerome     | Peng      |
(2 rows)
```

也可以对缺失的子节点设置默认值。

```
select decoded.*
from
    test_people,
    xmltable(
        '//people/person'
        passing xml_data
        columns
            first_name text,
            last_name text,
            nick text DEFAULT '---'
    ) as decoded;
```

查询结果如下。在 nick 缺失时,使用默认值 "---" 代替。

```
 first_name | last_name | nick
------------+-----------+------
 Pavel      | Feng      | john
 Jerome     | Peng      | ---
(2 rows)
```

也可以对结果集添加首列行号。

```
select decoded.*
from
    test_people,
    xmltable(
        '//people/person'
        passing xml_data
        columns
            id for ordinality,
            first_name text,
            last_name text,
            nick text DEFAULT '---'
    ) as decoded;
```

查询结果如下。字段值为添加的行号。

```
 id | first_name | last_name | nick
----+------------+-----------+------
  1 | Pavel      | Feng      | john
  2 | Jerome     | Peng      | ---
```

(2 rows)

还可以使用 XPath 函数进行处理，如组合 first_name 与 last_name。

```
select decoded.*
from
   test_people,
   xmltable(
      '//people/person'
      passing xml_data
      columns
         id for ordinality,
         full_name text PATH 'concat(first_name, " ", last_name)',
         nick text DEFAULT '---'
   ) as decoded;
```

查询结果如下。使用 concat 函数对 first_name 与 last_name 进行组合。

```
id | full_name   | nick
---+-------------+------
 1 | Pavel Feng  | john
 2 | Jerome Peng | ---
(2 rows)
```

1.5 PostgreSQL 10 的系统层变化

PostgreSQL 10 的系统层变化主要体现在 XLOG 重命名、系统元数据引入、配置参数引入、口令加密安全性提高、预置角色变化、附加模块变化。

1.5.1 XLOG 重命名

在 PostgreSQL 10 中，对 XLOG 及 CLOG 相关的工具、数据目录、系统函数、配置参数日志目录进行了重命名，如表 1-2～表 1-5 所示。这样做可以避免用户把相关文件当成普通的日志文件。

表 1-2 XLOG 及 CLOG 相关的工具重命名

PostgreSQL 9.6	PostgreSQL 10
pg_receivexlog	pg_receivewal
pg_resetxlog	pg_resetwal
pg_xlogdump	pg_waldump

表 1-3 数据目录重命名

PostgreSQL 9.6	PostgreSQL 10
pg_clog	pg_xact
pg_xlog	pg_wal

表 1-4　系统函数重命名

PostgreSQL 9.6	PostgreSQL 10
pg_current_xlog_flush_location	pg_current_wal_flush_lsn
pg_current_xlog_insert_location	pg_current_wal_insert_lsn
pg_current_xlog_location	pg_current_wal_lsn
pg_is_xlog_replay_paused	pg_is_wal_replay_paused
pg_last_xlog_receive_location	pg_last_wal_receive_lsn
pg_last_xlog_replay_location	pg_last_wal_replay_lsn
pg_switch_xlog	pg_switch_wal
pg_xlogfile_name	pg_walfile_name
pg_xlogfile_name_offset	pg_walfile_name_offset
pg_xlog_location_diff	pg_wal_lsn_diff
pg_xlog_replay_pause	pg_wal_replay_pause
pg_xlog_replay_resume	pg_wal_replay_resume

表 1-5　配置参数日志目录重命名

PostgreSQL 9.6	PostgreSQL 10
pg_log	log

1.5.2　系统元数据引入

PostgreSQL 10 新引入了 11 个系统元数据，如表 1-6 所示。

表 1-6　PostgreSQL 10 新引入的系统元数据

元数据名称	描述
pg_hba_file_rules	显示 pg_hba.conf 文件信息
pg_partitioned_table	显示分区表信息
pg_publication	显示逻辑复制发布信息
pg_publication_rel	显示逻辑复制发布与发布对象之间的映射信息
pg_publication_tables	显示逻辑复制发布与表之间的映射信息，与 pg_publication_rel 视图有所不同的是，每个具体的表都有一行记录
pg_sequence	显示序列的静态元数据信息，从序列移动到该系统表，以便同时查询多个序列
pg_sequences	显示序列的元数据信息，包括静态信息和动态信息，可以更加方便地查询序列的当前值，而不需要使用 nextval 函数去推进
pg_stat_subscription	显示逻辑复制订阅的状态信息
pg_statistic_ext	显示由 create statistics 命令创建的扩展统计信息
pg_subscription	显示逻辑复制订阅信息
pg_subscription_rel	显示逻辑复制订阅与订阅对象之间的映射信息

1.5.3 配置参数引入

PostgreSQL 10 新引入了 10 个配置参数，如表 1-7 所示。

表 1-7　PostgreSQL 10 新引入的配置参数

参数名称	描述
enable_gathermerge	启用或禁用 query planner 对 Gather Merge 的使用
max_logical_replication_workers	设置 logical replication workers 进程的最大数量
max_parallel_workers	设置 parallel workers 进程的最大数量
max_pred_locks_per_page	设置控制在将锁提升为覆盖整页之前，单页中能有多少个元组被 predicate locked
max_pred_locks_per_relation	设置控制在将锁提升为覆盖整个对象之前，单个对象中能有多少个面或元组被 predicate locked
max_sync_workers_per_subscription	设置控制在订阅初始化或新表加入时的 initial data copy 的并行度
min_parallel_index_scan_size	设置考虑使用 Parallel Scan 的索引数据的最小值
min_parallel_table_scan_size	设置考虑使用 Parallel Scan 的表数据的最小值
ssl_dh_params_file	设置支持用户自定义参数 openssl dh，规避了 PostgreSQL 10 之前版本的硬编码
wal_consistency_checking	设置 WAL 资源管理器检测 WAL 一致性的粒度

1.5.4 口令加密安全性提高

很多用户习惯使用 MD5 方式连接数据库来进行安全管理，但是目前使用 MD5 方式已经不再安全，这就需要使用新的身份验证方式。PostgreSQL 10 开始支持 SCRAM-SHA-256 方式，使用这种方式比使用 MD5 方式更安全。目前，虽然 MD5 方式仍然可以被使用，但是强烈建议使用 SCRAM-SHA-256 方式。

1.5.5 预置角色变化

从 PostgreSQL 9.6 开始，系统支持预置角色对需要经常使用的功能进行权限访问。管理员可以通过 GRANT 命令把预置角色授予用户，让这些用户能够访问指定的功能。

PostgreSQL 9.6 提供了第一个预置角色 pg_signal_backend，PostgreSQL 10 新增了 4 个与监控相关的预置角色，以便让数据库的权限分组管理更加便捷。

PostgreSQL 10 支持的预置角色如表 1-8 所示。

表 1-8　PostgreSQL 10 支持的预置角色

预置角色名称	描述
pg_read_all_settings	可以读取所有配置变量，包括那些通常只对超级用户可见的信息
pg_read_all_stats	可以读取所有 pg_stat_* 视图信息，以及相关的扩展统计信息，包括那些通常只对超级用户可见的信息

续表

预置角色名称	描述
pg_stat_scan_tables	可以执行一些监控函数，需要获取表的 Access Share 锁
pg_monitor	可以执行各种监视视图和函数，是 pg_read_all_settings、pg_read_all_stats 和 pg_stat_scan_tables 的成员
pg_signal_backend	可以使用 pg_cancel_backend 函数或 pg_terminate_backend 函数对后端进程发送信号、取消后端进程查询或中止后端进程

1.5.6 附加模块变化

PostgreSQL 10 中附加模块的变化具体如下。
- postgres_fdw 模块外部表下推特性增强。
- file_fdw 模块增加了 program 选项，可以通过执行程序来读取外部数据。
- 新增 amcheck 模块，用于检验索引的一致性和完整性。
- btree_gist 模块的 GiST 索引新增对 UUID 数据类型和枚举数据类型的支持。
- btree_gin 模块的 GIN 索引新增对枚举数据类型的支持。

1.6 本章小结

　　PostgreSQL 10 是一个非常重要的版本。关于版本号，由三段（X、Y、Z）数字变为了两段（X、Y）数字。在系统层变化方面，对 XLOG 及 CLOG 相关的工具、数据目录、系统函数、配置参数日志目录进行了重命名，以避免用户把相关文件当成普通的日志文件，XLOG 及 CLOG 相关的工具、数据目录、系统函数、配置参数日志目录需要注意切换。在开发易用性提升方面，引入了声明式分区及扩展了统计信息。在搭建主备环境方面，充分利用了仲裁提交使用多个同步设备来提高可靠性。此外，可以放心使用 HASH 索引。在运维管理优化方面，需要重点关注活动会话视图新增的 backend_type 字段，其可以用于监控系统后台进程及筛选用途。同时，结合新增的预置角色可以更好地进行最小化权限控制。后台脚本编写人员可以使用 PSQL 工具引入的条件式交互语法来优化代码结构。开发人员可以充分关注声明式分区和逻辑复制的新特性，以及标识列对比序列的优化。全文检索对 JSON 数据类型的支持和 xmltable 函数对 XML 格式数据的解析处理在相应的场景下也具有不容忽视的作用。

第 2 章

PostgreSQL 11 新特性

2.1 PostgreSQL 11 的主要性能提升

与 PostgreSQL 10 相比,PostgreSQL 11 的性能有了很大的提升,特别是在大数据库集中和高计算负载的情况下,主要体现在 WAL 可配置、B-Tree 索引并行创建、HASH 操作及 HASH JOIN 操作支持并行、其他并行特性支持、表达式索引引入、覆盖索引引入、实时编译引入、缓存管理改进、UPDATE 操作和 DELETE 操作支持下推。

2.1.1 WAL 可配置

PostgreSQL 自发布以来,单个 WAL 文件的大小始终是 16MB。最开始,它甚至是一个可编译限制项,后来改为编译时的选项。从 PostgreSQL 11 开始,这些 WAL 文件的大小可以在创建实例时更改,这为管理员提供了额外配置和优化 PostgreSQL 的方法。

那么 WAL 文件的大小对数据库的性能有什么影响呢?如果正在运行一个 99%都是读负载的数据库应用,或写负载的数据库应用,那么负载为零或接近零。只有在运行一个写密集型工作负载的数据库应用时,才会看到效果,这样改变 WAL 文件的大小才会物有所值。

下面说明在初始化数据库时如何配置 WAL 文件的大小。

```
$ initdb -D $PGDATA --wal-segsize=64
```

在初始化数据库时,配置 WAL 文件的大小为 64MB,在启动数据库后可以确认该文件的大小为 64MB。

```
$ ls -lh $PGDATA/pg_wal
total 64M
-rw------- 1 postgres postgres  64M Oct 16 16:03 000000010000000000000001
```

```
drwx------ 2 postgres postgres    6 Oct 16 15:56 archive_status
```

2.1.2 B-Tree 索引并行创建

在创建索引时，数据库通常使用 CPU 来完成工作。随着数据量的持续增长，创建索引开始成为一个问题。PostgreSQL 11 支持并行创建 B-Tree 索引，并可以通过维护操作并行度参数 max_parallel_maintenance_workers 来设置最大允许的并行进程数。在数据迁移的一些场景下，利用 CPU 的多核特性并行创建索引可以显著加快速度。

下面在笔记本虚拟机中测试并行创建 B-Tree 索引的特性。

创建一个测试表，插入 3000 万条数据。

```
create table tab_big(user_id int4,user_name text,ctime timestamp(6) without time
    zone default clock_timestamp());

insert into tab_big(user_id,user_name) select n ,n || '_data' from
    generate_series(1,30000000) n;
```

上面的语句在数据库执行之后的非并行模式下创建索引，观察执行时间。

```
postgres=# set max_parallel_maintenance_workers =0;
SET
Time: 0.290 ms

postgres=# create index on tab_big(ctime);
CREATE INDEX
Time: 169818.373 ms (02:49.818)
```

设置参数 max_parallel_maintenance_workers 的值为 0。在非并行模式下创建索引大约需要 170s。

下面设置并行度为 2，在并行模式下创建索引大约需要 76s，比在非并行模式下创建索引节省了一倍以上的时间。

```
postgres=# set max_parallel_maintenance_workers =2;
SET
Time: 0.290 ms

postgres=# create index on tab_big(ctime);
CREATE INDEX
Time: 75534.183 ms (01:15.534)
```

在并行模式下，PostgreSQL 11 会通过并行度设置独立创建的后台 worker 进程来提高效率。通过下面的代码可以查询到新增的 parallel worker 进程。

```
postgres=# select pid,query,backend_type from pg_stat_activity where query like
    'create index%';
 pid  |           query            |   backend_type
------+----------------------------+------------------
 3936 | create index on tab_big(ctime); | client backend
 4016 | create index on tab_big(ctime); | parallel worker
(2 rows)
```

2.1.3　HASH 操作及 HASH JOIN 操作支持并行

大多数等值连接查询的 SQL 语句都在后台进行 HASH JOIN 操作，在 PostgreSQL 11 中 HASH 操作及 HASH JOIN 操作支持并行，可以充分利用硬件的能力来提高性能。并行 HASH 操作及 HASH JOIN 操作受参数 enable_parallel_hash 的控制，默认值为 on。

下面测试并行 HASH 操作及 HASH JOIN 操作的特性。

先创建一个大表，插入 5000 万条数据，再创建一个小表，插入 800 万条数据。

```
create table t_big(id int4,name text,create_time timestamp without time zone);
insert into t_big(id,name,create_time)select n, n|| '_test',clock_timestamp() from
    generate_series(1,50000000) n ;

create table t_small(id int4, name text);

insert into t_small(id,name)select n, n|| '_small' from generate_series(1,8000000) n ;
```

关闭并行 HASH 操作及 HASH JOIN 操作的特性，设置参数 enable_parallel_hash 的值为 off，查看真实执行计划。

```
postgres=# set enable_parallel_hash = off;
SET

postgres=# explain analyze select t_small.name  from t_big join t_small on
    (t_big.id = t_small.id) and t_small.id < 100;
                        QUERY PLAN
-------------------------------------------------------------------------------
 Gather  (cost=151869.66..690486.34 rows=800 width=13) (actual
   time=996.137..3496.940 rows=99 loops=1)
   Workers Planned: 4
   Workers Launched: 4
   ->  Hash Join  (cost=150869.66..689406.34 rows=200 width=13) (actual
   time=2990.847..3490.557 rows=20 loops=5)
         Hash Cond: (t_big.id = t_small.id)
         ->  Parallel Seq Scan on t_big  (cost=0.00..491660.86 rows=12499686 width=4)
  (actual time=0.240..1392.062 rows=10000000 loops=5)
         ->  Hash  (cost=150859.66..150859.66 rows=800 width=17) (actual
   time=890.943..890.943 rows=99 loops=5)
               Buckets: 1024  Batches: 1  Memory Usage: 13kB
               ->  Seq Scan on t_small  (cost=0.00..150859.66 rows=800 width=17)
  (actual time=884.288..890.906 rows=99 loops=5)
                     Filter: (id < 100)
                     Rows Removed by Filter: 7999901
 Planning Time: 0.154 ms
 Execution Time: 3496.982 ms
(13 rows)
```

打开并行 HASH 操作及 HASH JOIN 操作的特性，设置参数 enable_parallel_hash 的值为 on，再次查看真实执行计划。

```
postgres=# set enable_parallel_hash = on;
```

```
SET
postgres=# explain analyze select t_small.name from t_big join t_small on
   (t_big.id = t_small.id) and t_small.id < 100;
                           QUERY PLAN
-------------------------------------------------------------------------------
 Gather  (cost=76862.42..615477.60 rows=800 width=13) (actual time=197.399..2738.010
   rows=99 loops=1)
   Workers Planned: 4
   Workers Launched: 4
   ->  Parallel Hash Join  (cost=75862.42..614397.60 rows=200 width=13) (actual
 time=2222.347..2729.943 rows=20 loops=5)
         Hash Cond: (t_big.id = t_small.id)
         ->  Parallel Seq Scan on t_big  (cost=0.00..491660.86 rows=12499686 width=4)
 (actual time=0.038..1330.836 rows=10000000 loops=5)
         ->  Parallel Hash  (cost=75859.92..75859.92 rows=200 width=17) (actual
 time=191.484..191.484 rows=20 loops=5)
               Buckets: 1024  Batches: 1  Memory Usage: 40kB
               ->  Parallel Seq Scan on t_small  (cost=0.00..75859.92 rows=200
 width=17) (actual time=152.436..191.385 rows=20 loops=5)
                     Filter: (id < 100)
                     Rows Removed by Filter: 1599980
 Planning Time: 0.183 ms
 Execution Time: 2738.068 ms
(13 rows)
```

对比上面两个执行计划可以发现，打开并行 HASH 操作及 HASH JOIN 操作的特性之后使用了 PARALLEL HASH JOIN 及 PARALLEL HASH，打开并行 HASH 操作及 HASH JOIN 操作的特性之前使用了 HASH 及 HASH JOIN。打开并行 HASH 操作及 HASH JOIN 操作的特性之后的执行时间大约为 2.7s，比打开并行 HASH 操作及 HASH JOIN 操作的特性之前的执行时间（大约为 3.4s）提高了 20%以上。

2.1.4　其他并行特性支持

PostgreSQL 11 支持的其他并行特性如下。
- CREATE TABLE AS SELECT。
- CREATE MATERIALIZED VIEW AS SELECT。
- SELECT INTO。

并行 CREATE TABLE AS SELECT 操作的示例如下。

```
postgres=# explain create table tab_cnt as select count(*) from tab_big ;
                           QUERY PLAN
-------------------------------------------------------------------------------
 Finalize Aggregate  (cost=376855.22..376855.23 rows=1 width=8)
   ->  Gather  (cost=376855.00..376855.21 rows=2 width=8)
         Workers Planned: 2
         ->  Partial Aggregate  (cost=375855.00..375855.01 rows=1 width=8)
```

```
                  -> Parallel Seq Scan on tab_big  (cost=0.00..344605.00 rows=12500000
 width=0)
(5 rows)
```

并行 CREATE MATERIALIZED VIEW AS SELECT 操作的示例如下。

```
postgres=# explain create materialized view mv1 as select count(*) from tab_big ;
                                    QUERY PLAN
-----------------------------------------------------------------------------------
 Finalize Aggregate  (cost=376855.22..376855.23 rows=1 width=8)
   -> Gather  (cost=376855.00..376855.21 rows=2 width=8)
         Workers Planned: 2
         -> Partial Aggregate  (cost=375855.00..375855.01 rows=1 width=8)
               -> Parallel Seq Scan on tab_big  (cost=0.00..344605.00 rows=12500000
 width=0)
(5 rows)
```

并行 SELECT INTO 操作的示例如下。

```
postgres=# explain select count(*) into val from tab_big ;
                                    QUERY PLAN
-----------------------------------------------------------------------------------
 Finalize Aggregate  (cost=376855.22..376855.23 rows=1 width=8)
   -> Gather  (cost=376855.00..376855.21 rows=2 width=8)
         Workers Planned: 2
         -> Partial Aggregate  (cost=375855.00..375855.01 rows=1 width=8)
               -> Parallel Seq Scan on tab_big  (cost=0.00..344605.00 rows=12500000
 width=0)
(5 rows)
```

2.1.5 表达式索引引入

PostgreSQL 对于普通的查询，通过查看内部统计信息来优化。表的每列都有统计信息，PostgreSQL 根据统计信息估算查询列返回结果集的数量，如果数量很少，那么 PostgreSQL 会考虑使用索引查询。

如果查询列为虚拟列或表达式，如将(column1+column2)这种基于表达式运算的虚拟列作为查询条件，那么 PostgreSQL 11 可以使用表达式索引，从而创建更好的执行计划，这可以为一些应用场景提高效率做出贡献。

下面查询创建时间在两天内的热数据。

创建表及准备数据。

```
create table foo(id int,dt timestamp);

insert into foo select f2,f1 from generate_series('2018-10-18'::timestamp,
now(),'1 hours') with ordinality as s(f1,f2);
```

根据查询列创建时间，查询两天内的热数据。

```
select * from foo where (dt + (interval '2 days')) < now();
```

如果对查询列创建索引，那么执行计划如下。

```
postgres=# create index on foo(dt);
CREATE INDEX
```

```
postgres=# explain select * from foo where (dt + (interval '2 days')) < now();
                        QUERY PLAN
---------------------------------------------------------------
 Seq Scan on foo  (cost=0.00..762.73 rows=11100 width=12)
   Filter: ((dt + '2 days'::interval) < now())
(2 rows)
```

可以发现，直接对查询列创建索引，PostgreSQL 11 并不会使用索引查询。

如果对查询表达式创建索引，那么执行计划如下。

```
postgres=# create index on foo((dt + (interval '2 days')));
CREATE INDEX

postgres=# explain select * from foo where (dt + (interval '2 days')) < now();
                            QUERY PLAN
-----------------------------------------------------------------------
 Bitmap Heap Scan on foo  (cost=214.32..588.57 rows=11100 width=12)
   Recheck Cond: ((dt + '2 days'::interval) < now())
   ->  Bitmap Index Scan on foo_expr_idx  (cost=0.00..211.54 rows=11100 width=0)
         Index Cond: ((dt + '2 days'::interval) < now())
(4 rows)
```

可以发现，使用新的表达式索引之后，PostgreSQL 11 能使用 Bitmap Index Scan 来查询数据。

2.1.6 覆盖索引引入

使用关键字 include 创建覆盖索引可以让更多场景使用 Index Only Scan，另外复合多列索引的列如果不支持相应操作分类符，那么可以使用关键字 include 创建索引。

下面是使用覆盖索引的第一种场景：覆盖索引可以让更多场景使用 Index Only Scan。

创建测试表，插入 100 万条数据，并收集表的统计信息。

```
create table t_include1(id int4 primary key, name text);

insert into t_include1(id,name)
select n,'test '||n from generate_series(1,1000000) n;

analyze t_include1;
```

由于 where 条件的 id 字段有索引，并且查询结果集只返回 id 字段，因此下面的语句中可以使用 Index Only Scan 直接从索引中获取数据，而不需要从磁盘中读取数据。

```
postgres=# explain select id from t_include1 where id<5;
                            QUERY PLAN
---------------------------------------------------------------------------
 Index Only Scan using t_include1_pkey on t_include1  (cost=0.42..8.50 rows=4 width=4)
   Index Cond: (id < 5)
(2 rows)
```

如果查询结果集需要增加 name 字段，同时返回 id 字段和 name 字段，那么 PostgreSQL 11 虽然可以使用 Index Scan，但还需要通过扫描堆表获取 name 字段，不能使用 Index Only Scan。

```
postgres=# explain select id,name from t_include1 where id<5;
                              QUERY PLAN
-----------------------------------------------------------------------
 Index Scan using t_include1_pkey on t_include1  (cost=0.42..8.50 rows=4 width=15)
   Index Cond: (id < 5)
(2 rows)
```

下面使用关键字 include 创建覆盖索引。这样既能保证 id 字段的唯一约束，又能在查询其他列时使用 Index Only Scan。

```
postgres=# create unique index idx_t_include1_inc on t_include1 (id) include
    (name);
CREATE INDEX
```

对表进行 VACUUM 操作及 ANALYZE 操作。

```
postgres=# vacuum analyze t_include1 ;
VACUUM
```

再次查询可以发现，使用覆盖索引后，该语句也可以使用 Index Only Scan 直接从索引中获取数据。

```
postgres=# explain select id,name from t_include1 where id<5;
                              QUERY PLAN
-----------------------------------------------------------------------
 Index Only Scan using idx_t_include1_inc on t_include1  (cost=0.42..4.51 rows=5
   width=15)
   Index Cond: (id < 5)
(2 rows)
```

使用关键字 include 创建覆盖索引，既可以避免多列复合索引可能发生的膨胀问题，又可以让索引更加轻量，同时也不用担心唯一约束的问题。注意，这个特性并不是无懈可击的，随着添加越来越多的数据到索引中，索引值会变得非常大，进而会产生一些问题。

下面是使用覆盖索引的第二种场景：不支持相应操作分类符的复合索引列可以使用关键字 include 创建覆盖索引。

创建测试表，id 字段是整型，user_info 字段是 JSON 数据类型。

```
create table t_include2(id int4, user_info json);
```

对 id 字段和 user_info 字段创建覆盖索引。

```
postgres=# create index  on t_include2 (id,user_info);
ERROR:  data type json has no default operator class for access method "btree"
HINT:  You must specify an operator class for the index or define a default
    operator class for the data type.
```

可以看到，由于 JSON 数据类型不支持默认的 B-Tree 索引因此创建失败。

在 PostgreSQL 11 中，可以使用关键字 include 创建覆盖索引。

```
postgres=# create index  on t_include2 (id) include(user_info);
CREATE INDEX
```

2.1.7 实时编译引入

实时编译（Just-In-Time，JIT）是 PostgreSQL 11 的一个重量级新特性。JIT 使用项

目编译器 LLVM 的架构来提升在 where 条件、指定列表、聚合，以及一些内部操作表达式中的编译速度。要使用 JIT，应先编译安装 LLVM，再编译安装 PostgreSQL，设置 --with-llvm 选项（可以使用 pg_config 工具来验证），最后在数据库中设置参数 jit 的值为 on。

2.1.8 缓存管理改进

PostgreSQL 11 为用户提供了更好的方法来管理共享缓冲区，通过 pg_prewarm 模块周期性地把共享缓冲区的内容转存到 autoprewarm.blocks 文件中，并在数据库实例重启后自动预加载共享缓冲区的内容，以便具有更好的性能。

随着内存变得越来越廉价，服务器内存配置越来越高，使用 shared_buffers 管理的内存也较大，内存大的系统可以从这个特性的改进中受益。

2.1.9 UPDATE 操作和 DELETE 操作支持下推

PostgreSQL 11 对外部表的重大改进之一是 UPDATE 操作和 DELETE 操作支持下推。下面分别测试 UPDATE 操作和 DELETE 操作下推。

进行环境准备。

```
create extension postgres_fdw;

create server remote_server foreign data wrapper postgres_fdw options (host
    '127.0.0.1',port '1116',dbname 'postgres');

create user mapping for public server remote_server options (user 'postgres');

create table local_tab1(id int,info text);

insert into local_tab1 select n,'a'||n from generate_series(1,1000) n;

create table local_tab2 as select * from local_tab1;

create foreign table remote_tab1(id int,info text)
server remote_server options (table_name 'local_tab1');

create foreign table remote_tab2(id int,info text)
server remote_server options (table_name 'local_tab2');
```

UPDATE 操作下推示例如下。

```
postgres=# explain (verbose,costs off)
 update remote_tab1 t1
    set info = t2.info
    from remote_tab2 t2
  where t1.id = t2.id and t1.id between 800 and 900;
                         QUERY PLAN
-------------------------------------------------------------------------------
```

```
Update on public.remote_tab1 t1
  -> Foreign Update
     Remote SQL: UPDATE public.local_tab1 r1 SET info = r2.info FROM
  public.local_tab2 r2 WHERE ((r1.id = r2.id)) AND ((r1.id >= 800)) AND ((r1.id <=
  900))
(3 rows)
```

DELETE 操作下推示例如下。

```
postgres=# explain (analyze,verbose,costs off)
 delete from remote_tab2
  using remote_tab1
  where remote_tab2.id = remote_tab1.id
and remote_tab2.id % 10 = 2 ;
                         QUERY PLAN
-----------------------------------------------------------------------
Delete on public.remote_tab2 (actual time=1.013..1.014 rows=0 loops=1)
  -> Foreign Delete (actual time=1.012..1.012 rows=0 loops=1)
     Remote SQL: DELETE FROM public.local_tab2 r1 USING public.local_tab1 r2
  WHERE ((r1.id = r2.id)) AND (((r1.id % 10) = 2))
Planning Time: 0.276 ms
Execution Time: 1.205 ms
(5 rows)
```

2.2 PostgreSQL 11 的可靠性提高

PostgreSQL 11 的可靠性提高主要体现在数据块校验和检测、B-Tree 索引坏块检测、查询 ID 由 32 位扩充为 64 位。

2.2.1 数据块校验和检测

PostgreSQL 11 新增了 pg_verify_checksums 工具，用于检测数据块校验和。要检测数据块校验和，数据库在初始化时需要打开校验和开关（-k,--data-checksums），在执行操作时需要关闭数据库实例，不支持在线操作。

在检测数据块校验和时，可以对整个 PGDATA 操作，也可以对单个对象操作。对单个对象进行检测的示例如下。

```
$ pg_verify_checksums -D $PGDATA -r 16409
pg_verify_checksums: checksum verification failed in file "data/base/13287/16409",
   block 0: calculated checksum 3A41 but expected 62AC Checksum scan completed
Data checksum version: 1
Files scanned: 1
Blocks scanned: 1
Bad checksums: 1
```

2.2.2 B-Tree 索引坏块检测

PostgreSQL 11 对 B-Tree 索引支持通过 amcheck 模块进行坏块检测，演示示例如下。
foo 表的 id 字段建立名为 foo_id_idx 的 B-Tree 索引。

```
postgres=# \d foo
                Table "public.foo"
 Column |            Type             | Collation | Nullable | Default
--------+-----------------------------+-----------+----------+---------
 id     | integer                     |           |          |
 dt     | timestamp without time zone |           |          |
Indexes:
    "foo_id_idx" btree (id)

postgres=# select pg_relation_filepath('foo_id_idx');
 pg_relation_filepath
----------------------
 base/13287/49164
(1 row)
```

使用 dd 命令模拟索引文件被损坏。

```
$ dd if=/dev/zero of=49164 bs=1 count=100
100+0 records in
100+0 records out
100 bytes (100 B) copied, 0.000801016 s, 125 kB/s
```

通过 amcheck 模块的 bt_index_check 函数进行检测。

```
postgres=# create extension amcheck;
CREATE EXTENSION

postgres=# select bt_index_check('foo_id_idx') ;
ERROR:  index "foo_id_idx" contains unexpected zero page at block 3
HINT:  Please REINDEX it.
```

使用 REINDEX 命令重建索引后，再次进行检测，没有报错信息。

```
postgres=# reindex index foo_id_idx ;
REINDEX

postgres=# select bt_index_check('foo_id_idx') ;
 bt_index_check
----------------

(1 row)
```

2.2.3 查询 ID 由 32 位扩充为 64 位

pg_stat_statements 模块是深入理解 PostgreSQL 11 性能的黄金标准工具。随着 PostgreSQL 11 服务的启动，pg_stat_statements 模块就会加载参数 shared_preload_libraries 并汇总有关服务器上运行的查询统计信息。

pg_stat_statements 模块提供了一个名为 queryid 的查询 ID，其长度一直是 32 位。在某些情况下键值可能会发生冲突。引入一个 64 位长度的查询 ID，使得发生冲突的概率变得极小，这是一个重大改进。

2.3 PostgreSQL 11 的运维管理优化

PostgreSQL 11 的运维管理优化主要体现在快速添加列、维护操作支持多个表、分区数据支持通过父表加载、新增 KILL 信号、WAL 支持离线重构、PSQL 工具支持记录语句执行情况。

2.3.1 快速添加列

ALTER TABLE 命令用于更改表的定义。在 PostgreSQL 11 中，ALTER TABLE 命令的执行情况得到了显著改善。以下示例显示了如何将列添加到表中，以及 PostgreSQL 11 如何处理这些新列。

```
ALTER TABLE x ADD COLUMN y int;
ALTER TABLE x ADD COLUMN z int DEFAULT 100;
```

上述第一行代码运行得很快，原因是在 PostgreSQL 11 中，列的默认值为 NULL。由于 PostgreSQL 所做的就是在系统目录中添加一列而不实际存储，不需要重写表，只需要更新数据字典，因此 DDL 能被瞬间执行。即使向一个 10 TB 大小的表中添加列，操作速度也会非常快，这是因为不必更新磁盘上的行。

上述第二行代码的情况则完全不同。DEFAULT 100 表示在行中添加了实际数据，在 PostgreSQL 10 及更早版本中，这意味着数据库必须重写整个表以添加这个新默认值。表越大，执行时间越长，如果表中包含数十亿行记录，那么无法将其锁定并重写。

在实际生产环境下给大表添加带默认值的字段非常困难，通常分两步进行。第一步，先添加不带默认值的字段。第二步，编写函数批量刷新新增字段的默认值。第二步比较麻烦，可以在业务低谷或申请停服窗口一次性地完成新增带默认值字段的操作。

从 PostgreSQL 11 开始，可以在不重写整个表的情况下向表中添加固定的默认值，这样大大减轻了更改数据结构的负担。

下面分别在 PostgreSQL 10 和 PostgreSQL 11 中进行测试。

创建测试表并插入 1000 万条数据。

```
CREATE TABLE t1(id int4, name text);

INSERT INTO t1 (id,name ) SELECT n, n || '_ALTER TABLE TEST ' FROM generate_series
   (1,10000000) n;

ANALYZE t1;
```

在 PostgreSQL 10 中查看表的物理文件号。

```
postgres=# SELECT relname,relfilenode FROM pg_class WHERE relname='t1';
```

```
 relname | relfilenode
---------+-------------
 t1      |       25672
(1 row)
```

在 PostgreSQL 10 中新增带默认值的非空字段。

```
postgres=# \timing
Timing is on.

postgres=# ALTER TABLE t1 ADD COLUMN flag text DEFAULT 'abcd';
ALTER TABLE
Time: 15540.002 ms (00:15.540)
```

执行时间较长，需要大约 16 秒。

```
postgres=# ANALYZE t1;
ANALYZE

postgres=# SELECT relname,relfilenode FROM pg_class WHERE relname='t1';
 relname | relfilenode
---------+-------------
 t1      |       25679
(1 row)
```

分析表后，再次查看表的物理文件号，会发现表的物理文件号有变化，说明表已被重写。

在 PostgreSQL 11 中进行同样的操作，查看表的物理文件号。

```
postgres=# SELECT relname,relfilenode FROM pg_class WHERE relname='t1';
 relname | relfilenode
---------+-------------
 t1      |       16802
(1 row)
```

在 PostgreSQL 11 中新增带默认值的非空字段。

```
postgres=# \timing
Timing is on.

postgres=# ALTER TABLE t1 ADD COLUMN flag text DEFAULT 'abcd';
ALTER TABLE
Time: 40.743 ms
```

执行时间只需要大约 41ms，瞬间完成。

```
postgres=> ANALYZE t1;
ANALYZE

postgres=# SELECT relname,relfilenode FROM pg_class WHERE relname='t1';
 relname | relfilenode
---------+-------------
 t1      |       16802
(1 row)
```

分析表后，再次查看表的物理文件号，会发现表的物理文件号没有变化，依然是 16802，说明表没有被重写。

2.3.2 维护操作支持多个表

从 PostgreSQL 11 开始，进行 VACUUM 操作及 ANALYZE 操作支持同时指定多个表。

```
postgres=# VACUUM data1, data2;
VACUUM

postgres=# ANALYZE data1, data2;
ANALYZE
```

2.3.3 分区数据支持通过父表加载

在通过 pg_dump 工具或 pg_dumpall 工具导出分区数据时，会通过 PostgreSQL 11 中新增的 --load-via-partition-root 选项使用父表而不是分区本身来备份数据，这对源数据库和目标数据库异构分区数据操作非常有用。

分区表备份示例一：只对需要的分区进行备份。
```
$ pg_dump -f dmp.sql -t 分区1 -t 分区2...
```
分区表备份示例二：备份到父表中，便于目标数据库重构分区策略。
```
$ pg_dump -f dmp.sql --load-via-partition-root -t 分区1 -t 分区2 ...
```

2.3.4 新增 KILL 信号

在有些场景下，客户端进程无法通过 pg_terminate_backend 函数被中止，通常使用 kill 子命令杀死进程，此时不会做任何处理，不会释放相关资源。如果用户后台进程正在修改共享内存，那么共享内存可能会被损坏，实例可能会受到影响。PostgreSQL 11 对 pg_ctl 命令的 kill 子命令新增了 KILL 信号，对异常客户端可以使用如下命令来中止。
```
$ pg_ctl kill KILL $PID
```

2.3.5 WAL 支持离线重构

从 PostgreSQL 11 开始，WAL 文件的大小可以在实例初始化时设置，如果在实例初始化时忘记了设置，那么可以使用 pg_resetwal 工具来重新调整 WAL 文件。

下面演示如何重新调整 WAL 文件的大小。

查看当前数据库的 pg_wal 目录。
```
$ ll $PGDATA/pg_wal/
total 2.3G
-rw------- 1 postgres postgres 16M Sep 30 14:45 000000010000001700000013
-rw------- 1 postgres postgres 16M Sep 30 14:45 000000010000001700000014
-rw------- 1 postgres postgres 16M Sep 30 14:45 000000010000001700000015
-rw------- 1 postgres postgres 16M Sep 30 14:45 000000010000001700000016
-rw------- 1 postgres postgres 16M Sep 30 14:45 000000010000001700000017
-rw------- 1 postgres postgres 16M Sep 30 14:45 000000010000001700000018
-rw------- 1 postgres postgres 16M Sep 30 14:45 000000010000001700000019
-rw------- 1 postgres postgres 16M Sep 30 14:45 00000001000000170000001A
```

```
-rw------- 1 postgres postgres 16M Sep 30 14:45 00000001000000170000001B
...
drwx------ 2 postgres postgres 16K Oct 16 08:38 archive_status
```

pg_wal 目录中已有大量 WAL 文件，WAL 文件的大小为 16MB，现将 WAL 文件调整成 64MB。

在使用 pg_resetwal 工具调整 WAL 文件之前需要先关闭数据库。

```
$ pg_ctl stop -D $PGDATA
waiting for server to shut down... done
server stopped
```

在数据库关闭成功之后，使用 pg_resetwal 工具调整 WAL 文件的大小为 64MB。

```
$ pg_resetwal --wal-segsize=64 -D $PGDATA
```

下面验证 WAL 文件的大小。

```
$ ls -lh $PGDATA/pg_wal/
total 64M
-rw------- 1 postgres postgres 64M Oct 16 08:55 000000010000001700000029
drwx------ 2 postgres postgres 16K Oct 16 08:55 archive_status
```

可以发现，pg_wal 目录中原有的 WAL 文件已经被清理，同时生成了一个大小为 64MB 的新文件。

2.3.6　PSQL 工具支持记录语句执行情况

PostgreSQL 11 对 PSQL 工具的主要变化是新增 ERROR、SQLSTATE、ROW_COUNT、LAST_ERROR_MESSAGE、LAST_ERROR_SQLSTATE 这 5 个变量用来记录 SQL 语句的执行结果状态和错误信息。这些变量的值在 SQL 语句执行后刷新，主要用于在编写脚本时捕获 SQL 语句的执行结果。

下面通过示例演示几个变量的作用。

```
postgres=# select * from ;
ERROR:  syntax error at or near ";"
LINE 1: select * from ;
                      ^
Time: 0.207 ms

postgres=# \echo :ERROR
true

postgres=# \echo :SQLSTATE
42601

postgres=# \echo :ROW_COUNT
0
```

如果 SQL 语句执行出错，那么变量 ERROR 的值为 true，变量 SQLSTATE 返回报错代码，变量 ROW_COUNT 的值为 0。

下面的 SQL 语句执行成功之后，再次查看上面几个变量的值。

```
postgres=# SELECT * FROM generate_series(1,3);
 generate_series
```

```
----------------
           1
           2
           3
(3 rows)

postgres=# \echo :ERROR
false

postgres=# \echo :SQLSTATE
00000

postgres=# \echo :ROW_COUNT
3
```

可以看出，在 SQL 语句执行成功之后，变量 ERROR 的值为 false，变量 SQLSTATE 的值为 00000，变量 ROW_COUNT 的值为 3。

2.4　PostgreSQL 11 的开发易用性提升

PostgreSQL 11 的开发易用性提升主要体现在声明式分区增强、支持事务控制的存储过程、逻辑复制支持 TRUNCATE 操作、窗口函数增强。

2.4.1　声明式分区增强

PostgreSQL 10 支持两种声明式分区：范围分区及列表分区。PostgreSQL 11 新增支持 HASH 分区。HASH 分区根据分区键的 HASH 值进行分布式存储，分区键可以是单列的也可以是多列的。为了保持分区均匀，需要选择合适的分区键。

下面是对一个顾客信息表进行 HASH 分区的示例。

```
CREATE TABLE customer(
    cid    int4  PRIMARY KEY,
    cname  character varying(64),
    ctime  timestamp(6) without time zone
) PARTITION BY HASH(cid);
```

创建具体的分区。

```
CREATE TABLE customer_p0 PARTITION OF customer FOR VALUES WITH(MODULUS 4, REMAINDER
    0);
CREATE TABLE customer_p1 PARTITION OF customer FOR VALUES WITH(MODULUS 4, REMAINDER
    1);
CREATE TABLE customer_p2 PARTITION OF customer FOR VALUES WITH(MODULUS 4, REMAINDER
    2);
CREATE TABLE customer_p3 PARTITION OF customer FOR VALUES WITH(MODULUS 4, REMAINDER
    3);
```

HASH 分区包含两个属性，MODULUS 属性用于设置 HASH 分区的个数，对每个分区来讲，该属性的值是固定的，REMAINDER 属性用于设置 HASH 分区键对应取余的余数。

PostgreSQL 11 的分区功能还有不少亮点，首先是默认分区，如果现有分区都不匹配，那么可以创建默认分区。下面是它的工作原理。

```
postgres=# CREATE TABLE default_part PARTITION OF some_table DEFAULT;
CREATE TABLE
```

在上面的代码中，所有在任何分区条件都不匹配的行将最终出现在默认分区。

在 PostgreSQL 10 中，行不能（轻松地）从一个分区移动到另一个分区。假设有一个按照省/市区分的分区，如果一个人从北京市搬回湖北省，那么不能使用一个简单的 UPDATE 操作来实现，必须删除旧行并插入新行。在 PostgreSQL 11 中可以使用完全透明的方式将行从一个分区移动到另一个分区。

除此之外，为父表创建的索引将自动确保所有子表都被使用，这样可以降低索引被忘记创建的可能性。另外，可以添加全局唯一索引，分区表可以强制执行唯一约束。分区裁剪现在要快得多，PostgreSQL 11 能够智能处理跨分区关联，以及跨分区聚合的问题。

注意，表分区在 PostgreSQL 11 中有如下限制。
- 允许创建全局主键、唯一约束，但必须包含分区键。
- 允许创建全局外键，不允许创建排他约束。
- 允许创建全局索引，全局索引必须包含分区键。
- 普通表不允许外键引用分区表中的字段。
- 普通表不能继承分区父子表。
- 分区上的行级触发器不支持过渡表。

2.4.2 支持事务控制的存储过程

PostgreSQL 一直以来仅支持函数功能，没有存储过程。函数结构体的部分是一个单独的事务，而存储过程可以包含多个事务。从其他数据库迁移的数据库代码通常需要大量手动修改工作，这是因为存储过程可能包含带有开始、提交和回滚的事务。

创建存储过程。

```
CREATE [ OR REPLACE ] PROCEDURE
  name ( [ [ argmode ] [ argname ] argtype [ { DEFAULT | = } default_expr ]
  [, ...] ] )
{ LANGUAGE lang_name
  | TRANSFORM { FOR TYPE type_name } [, ... ]
  | [ EXTERNAL ] SECURITY INVOKER | [ EXTERNAL ] SECURITY DEFINER
  | SET configuration_parameter { TO value | = value | FROM CURRENT }
  | AS 'definition'
  | AS 'obj_file', 'link_symbol'
} ...
```

调用存储过程。

```
CALL name ( [ argument ] [, ...] )
```

以下存储过程代码说明了如何在一个存储过程中执行两个事务。

```
postgres=# CREATE PROCEDURE test_proc()
LANGUAGE plpgsql
```

```
AS $$
 BEGIN
   CREATE TABLE a (aid int);
   CREATE TABLE b (bid int);
   COMMIT;
   CREATE TABLE c (cid int);
   ROLLBACK;
 END;
$$;
CREATE PROCEDURE
```

注意，在中止第二个事务时，前两个语句已提交。稍后可以看到此更改的效果。

要运行该存储过程，可以使用关键字 CALL。

```
postgres=# CALL test_proc();
CALL
```

前两个表的创建语句已提交，由于存储过程内部的回滚，因此第三个表尚未创建。

```
postgres=# \d
 List of relations
 Schema | Name | Type  | Owner
--------+------+-------+--------
 public | a    | table | postgres
 public | b    | table | postgres
(2 rows)
```

存储过程中的另外一个特性是游标的 WITH HOLD 特性。通常，游标在事务结束后会自动关闭，不需要显式关闭，除非运行大事务，想要释放资源。使用游标的 WITH HOLD 特性虽可以突破事务范围的限制，但必须考虑提交对性能的影响，并且要记得手动关闭游标。

2.4.3 逻辑复制支持 TRUNCATE 操作

逻辑复制是从 PostgreSQL 10 开始出现的，可以在多个实例之间进行表的数据复制。在 PostgreSQL 10 中进行逻辑复制有如下限制：只支持 INSERT 操作、UPDATE 操作、DELETE 操作的数据复制，对表进行 TRUNCATE 操作、CREATE 操作、ALTER 操作时不会复制到远端。

表的 TRUNCATE 操作在 PostgreSQL 11 中得到了支持，可以期待其他限制在将来的版本中逐渐放开。

2.4.4 窗口函数增强

一直以来，PostgreSQL 是唯一支持大多数 SQL 标准窗口函数的开源数据库，但是对比 SQL 标准仍然存在一些小的功能缺失。窗口函数中的 Over 子句用于定义哪些行对窗口函数可见，PostgreSQL 全球社区在 PostgreSQL 11 中率先完全支持 SQL 标准 2011 的 Over 子句中的 frame unit groups 和 frame exclusion，任何其他主流数据库甚至是商业数据库都还不支持这些功能。

2.5 PostgreSQL 11 的系统层变化

PostgreSQL 11 的系统层变化主要体现在配置参数引入、预置角色变化、超级用户权限下放及附加模块变化。

2.5.1 配置参数引入

PostgreSQL 11 新引入了 21 个配置参数，如表 2-1 所示。

表 2-1　PostgreSQL 11 新引入的配置参数

参数名称	描述
data_directory_mode	PGDATA 对应的目录权限
enable_parallel_append	启用或禁用 query planner 对 parallel-aware append 的使用
enable_parallel_hash	启用或禁用 query planner 对 PARALLEL HASH JOIN 的使用
enable_partition_pruning	启用或禁用在 query planner 的分区表中取消分区
enable_partitionwise_aggregate	启用或禁用 query planner 对 partitionwise grouping or aggregation 的使用
enable_partitionwise_join	启用或禁用 query planner 对 partitionwise join 的使用，这允许分区表的 JOIN 在匹配的分区上执行 JOIN
jit	确定 JIT 是否被 PostgreSQL 11 使用
jit_above_cost	考虑使用 Parallel Scan 的表数据的最小大小
jit_debugging_support	注册 JIT 函数给 debugger 使用
jit_dump_bitcode	写出 LLVM 的 bitcode 以便于 JIT Debugging
jit_expressions	当 JIT 被启用时，本参数用于确定表达式是否被 JIT
jit_inline_above_cost	当 SQL 语句的查询成本高于本参数的值时，JIT 会尝试 inline functions and operators。这种优化虽然会增加计划时间，但是会加快执行速度
jit_optimize_above_cost	当 SQL 语句的查询成本高于本参数的值时，JIT 会尝试进行 expensive optimizations。这种优化虽然会增加计划时间，但是会加快执行速度
jit_profiling_support	如果 LLVM 具有所需的功能，那么会发出允许 perf 分析 JIT 生成的函数所需的数据
jit_provider	指定 JIT Provider Library 的名称
jit_tuple_deforming	当 JIT 被启用时，本参数用于确定是否允许 Tuple Deforming 进行 JIT
max_parallel_maintenance_workers	指定可以被单个系统维护命令启动的 parallel worker 进程的最大数量。CREATE INDEX、VACUUM 命令不带 full 选项

续表

参数名称	描述
parallel_leader_participation	允许 leader 进程在 Gather 和 Gather Merge node 下执行查询计划，而不是等待 worker 进程
ssl_passphrase_command	设置在需要获取用于解密 SSL 文件（私钥等）的密码短语时要调用的外部命令
ssl_passphrase_command_supports_reload	当一个 key file 需要 passphrase 时，本参数决定由参数 ssl_passphrase_command 设置的 passphrase command 在 configuration reload 期间是否被调用
vacuum_cleanup_index_scale_factor	VACUUM 有一个过程是根据死元组列表来删除索引元组，内部称 cleanup，如果目标索引是 B-Tree 索引，那么是否执行 cleanup 由本参数决定

2.5.2 预置角色变化

PostgreSQL 10 支持 5 个预置角色，而 PostgreSQL 11 则对 COPY 操作新增了 3 个预置角色，分别用于控制服务端文件的读写，以及执行程序的权限。

PostgreSQL 11 支持的预置角色如表 2-2 所示。

表 2-2　PostgreSQL 11 支持的预置角色

预置角色名称	描述
pg_execute_server_program	可以执行数据库服务端程序来配合 COPY 操作和其他允许执行服务端程序的函数
pg_read_server_files	可以使用 COPY 操作及其他文件访问函数在数据库服务端可访问的任意位置读取文件
pg_write_server_files	可以使用 COPY 操作及其他文件访问函数在数据库服务端可访问的任意位置写入文件
pg_read_all_settings	可以读取所有配置变量，包括那些通常只对超级用户可见的信息
pg_read_all_stats	可以读取所有 pg_stat_* 视图信息，以及相关的扩展统计信息，包括那些通常只对超级用户可见的信息
pg_stat_scan_tables	可以执行一些监控函数，需要获取表的 Access Share 锁
pg_monitor	可以执行各种监视视图和函数，是 pg_read_all_settings、pg_read_all_stats 和 pg_stat_scan_tables 的成员
pg_signal_backend	可以使用 pg_cancel_backend 函数或 pg_terminate_backend 函数对后端进程发送信号，取消后端进程查询或中止后端进程

2.5.3 超级用户权限下放

涉及数据库服务端文件读取的系统函数通常需要管理员权限，如 pg_ls_dir 等系统函数。PostgreSQL 11 支持少量文件读取的系统函数权限下放，可以使用 GRANT 命令将权限赋给普通用户，目前以下 4 个系统函数支持权限下放。
- pg_ls_dir。
- pg_read_file。
- pg_read_binary_file。
- pg_stat_file。

以上 4 个系统函数在 PostgreSQL 11 之前只有超级用户才有权限使用，从 PostgreSQL 11 开始 pg_rewind 工具可以不依赖超级用户而只需要分配以上 4 个系统函数权限。

2.5.4 附加模块变化

PostgreSQL 11 中附加模块的新特性包含一些原模块特性的增强，同时也包含一些新增特性，具体如下。
- adminpack 模块功能函数不再需要超级用户权限。
- btree_gin 模块的 GIN 索引新增对 BOOL、BPCHAR、UUID 数据类型的支持。

2.6 本章小结

PostgreSQL 11 对大数据库集和高计算负载进行了增强，主要包括声明式分区增强和 JIT 加速。PostgreSQL 11 对索引也进行了大量的改进，包括 B-Tree 索引并行创建、表达式索引引入、覆盖索引引入。在写密集型工作负载的场景下，用户可以在实例初始化时通过改变 WAL 文件的大小来提升性能，也可以在初始化后使用 pg_resetwal 工具离线重新调整 WAL 文件的大小。在运维管理优化方面，PostgreSQL 11 需要使用 pg_verify_checksums 工具，用于检测数据块校验和，COPY 操作对普通用户的权限应该进行读写限制，其他一些依赖超级用户权限的操作也不断被弱化。在开发易用性提升方面，PostgreSQL 11 可以兼容 Oracle 的存储过程，同时可以使用游标的 WITH HOLD 特性来突破事务范围的限制，但要记得手动关闭游标。另外，PostgreSQL 11 中大量的窗口函数新特性也值得探究。

第 3 章

PostgreSQL 12 新特性

3.1 PostgreSQL 12 的主要性能提升

与 PostgreSQL 12 之前的版本相比，PostgreSQL 12 的性能有了一定的提升，主要体现在 CTE 优化、索引效率提升、系统函数优化。

3.1.1 CTE 优化

CTE（Common Table Expressions）也就是 WITH 语句，用于把复杂查询分解为简短的片段，从而更易阅读和理解。使用 WITH 语句可以"物化"缓存多次重复的计算，减少冗余子查询数及减小函数的副作用。

在 PostgreSQL 12 之前的版本中需要注意 WITH 语句的常见缺陷：与普通查询相比，优化器无法将父查询的条件下推到外层查询中。

创建测试表并插入 200 万条数据，构造查询语句作为 WITH 语句。

```
with cte as (
   select * from foo
) select * from cte where id = 1;
```

在 PostgreSQL 11 中进行测试。

```
postgres=# explain analyze with cte as (
select * from foo
) select * from cte where id = 1;
                    QUERY PLAN
```

```
--------------------------------------------------------------------
CTE Scan on cte  (cost=36667.00..81667.00 rows=10000 width=36) (actual
   time=362.006..1497.847 rows=1 loops=1)
  Filter: (id = 1)
  Rows Removed by Filter: 1999999
  CTE cte
    -> Seq Scan on foo  (cost=0.00..36667.00 rows=2000000 width=36) (actual
   time=0.012..334.486 rows=2000000 loops=1)
 Planning Time: 0.804 ms
 Execution Time: 1783.474 ms
(7 rows)
```

从执行计划中可以看出，首先在 foo 表中进行全表扫描，其次进行 CTE Scan 并过滤 id 属性的值为 1 的记录，执行时间为 1783.474 ms，WITH 语句的性能较低。

在 PostgreSQL 12 中进行同样的测试。

```
postgres=# explain analyze with cte as (
select * from foo
) select * from cte where id = 1;
                          QUERY PLAN
--------------------------------------------------------------------
 Index Scan using foo_id_ix on foo  (cost=0.43..8.45 rows=1 width=37) (actual
   time=0.036..0.037 rows=1 loops=1)
   Index Cond: (id = 1)
 Planning Time: 0.110 ms
 Execution Time: 0.056 ms
(4 rows)
```

从执行计划中可以看出，WITH 语句中的条件下推到了外层查询中，这样直接使用了 Index Scan，执行时间降为 0.056 ms，大大提升了 WITH 语句的性能。

除此之外，在 PostgreSQL 12 中用户也可以手动设置是否物化。

```
postgres=# explain analyze with cte as not materialized (
select * from foo
) select * from cte where id = 1;
                          QUERY PLAN
--------------------------------------------------------------------
 CTE Scan on cte  (cost=36667.00..81667.00 rows=10000 width=36) (actual
   time=520.788..1695.881 rows=1 loops=1)
  Filter: (id = 1)
  Rows Removed by Filter: 1999999
  CTE cte
    -> Seq Scan on foo  (cost=0.00..36667.00 rows=2000000 width=37) (actual
   time=0.016..445.339 rows=2000000 loops=1)
 Planning Time: 0.081 ms
 Execution Time: 1711.266 ms
(7 rows)
```

可以看出，WITH 语句没有物化，这也是 PostgreSQL 12 默认的行为。

下面设置 WITH 语句物化后的行为与 PostgreSQL 11 一致。

```
postgres=# explain analyze with cte as materialized (
select * from foo
) select * from cte where id = 1;
```

```
                        QUERY PLAN
-----------------------------------------------------------------------
 CTE Scan on cte  (cost=36667.00..81667.00 rows=10000 width=36) (actual
   time=520.788..1695.881 rows=1 loops=1)
   Filter: (id = 1)
   Rows Removed by Filter: 1999999
   CTE cte
     -> Seq Scan on foo  (cost=0.00..36667.00 rows=2000000 width=37) (actual
   time=0.016..445.339 rows=2000000 loops=1)
 Planning Time: 0.081 ms
 Execution Time: 1711.266 ms
(7 rows)
```

PostgreSQL 12 的 WITH 语句默认不物化，查询条件可以下推到外层查询中，以避免中间产生结果数据，同时使用相关索引，能提高 WITH 语句的执行性能。

此外，使用 PostgreSQL 12 的优化器处理 WITH 语句更加智能，下面观察 PostgreSQL 11 的行为。

```
postgres=# explain with x as (select * from generate_series(1,5) as id)
select * from x;
                   QUERY PLAN
-----------------------------------------------------------------------
 CTE Scan on x  (cost=10.00..30.00 rows=1000 width=4)
   CTE x
     -> Function Scan on generate_series id  (cost=0.00..10.00 rows=1000 width=4)
(3 rows)
```

可以看出，预估函数返回 1000 行，这是不准确的。下面观察 PostgreSQL 12 的行为。

```
postgres=# explain with x as (select * from generate_series(1,5) as id)
select * from x;
                   QUERY PLAN
-----------------------------------------------------------------------
 Function Scan on generate_series id  (cost=0.00..0.05 rows=5 width=4)
(1 row)
```

可以看出有两点区别：一是 CTE Scan 消失了，这意味着 PostgreSQL 12 对 WITH 语句进行了内联并优化了 CTE Scan；二是预估的行数为 5，很明显，PostgreSQL 12 对 WITH 语句的优化效果更好。

3.1.2 索引效率提升

在 PostgreSQL 12 之前的版本中，B-Tree 索引重复的索引条目无序存储，会使得在插入时产生大量的开销，也会降低通过 VACUUM 操作回收整页的能力。在 PostgreSQL 12 中重复的索引条目按堆存储顺序排列，提高了 B-Tree 索引的性能和空间利用率。另外，创建 GiST、GIN 或 SP-GiST 索引产生的 WAL 消耗大大减少，这也会节省空间，提高磁盘的利用率。

3.1.3 系统函数优化

在 PostgreSQL 12 中，系统函数新增了 SUPPORT 属性。对于 SUPPORT 属性的函数，优化器可以准确地预估行。

下面使用 unnest 函数示例对比在 PostgreSQL 11 中和在 PostgreSQL 12 中的区别。

在 PostgreSQL 11 中，可以看出 Function Scan 预估的行数是 100 行，这其实并不准确。

```
postgres=# explain select * from unnest(array[1,2,3]);
                    QUERY PLAN
-----------------------------------------------------------
 Function Scan on unnest  (cost=0.00..1.00 rows=100 width=4)
(1 row)
```

在 PostgreSQL 12 中，Function Scan 预估的行数是准确的 3 行。

```
postgres=# explain select * from unnest(array[1,2,3]);
                    QUERY PLAN
-----------------------------------------------------------
 Function Scan on unnest  (cost=0.00..0.03 rows=3 width=4)
(1 row)
```

对比函数的定义可以看出，PostgreSQL 12 中的 unnest 函数新增了 SUPPORT 属性。

```
postgres=# \sf unnest(anyarray)
CREATE OR REPLACE FUNCTION pg_catalog.unnest(anyarray)
 RETURNS SETOF anyelement
 LANGUAGE internal
 IMMUTABLE PARALLEL SAFE STRICT ROWS 100 SUPPORT array_unnest_support
AS $function$array_unnest$function$
```

3.2 PostgreSQL 12 的运维管理优化

PostgreSQL 12 的运维管理优化主要体现在校验和开关控制、COPY FROM 命令数据过滤、用户级流复制超时控制、VACUUM 操作及 ANALYZE 操作锁跳过、表及索引清理解耦、索引在线重建、执行计划显示非默认参数、后台操作进度报告引入、备库升主库开放 SQL 接口、PSQL 工具帮助链接添加。

3.2.1 校验和开关控制

在 PostgreSQL 11 中如果数据库初始化时打开了校验和开关（-k,--data-checksums），那么可以在使用 pg_verify_checksums 工具关闭数据库实例之后，检测数据文件中的页是否已经被损坏。在 PostgreSQL 12 中 pg_verify_checksums 工具被重命名为 pg_checksums，同时新增了启用校验和与关闭校验和的开关项。

在使用 pg_checksums 工具操作之前需要先关闭数据库服务。

```
$ pg_ctl stop -D $PGDATA
```

启用校验和应使用-e,--enable 选项。在启用校验和时，每个数据文件都会被原地重写。

```
$ pg_checksums --enable -D $PGDATA
Checksum operation completed
Files scanned:  958
Blocks scanned: 3252
pg_checksums: syncing data directory
pg_checksums: updating control file
Checksums enabled in cluster
```

检测校验和应使用-c,--check 选项。如果没有检测到校验和错误，那么退出状态为零，如果检测到至少一个校验和错误，那么退出状态为非零。

```
$ pg_checksums --check -D $PGDATA
Checksum operation completed
Files scanned:  958
Blocks scanned: 3252
Bad checksums:  0
Data checksum version: 1
```

关闭校验和应使用-d,--disable 选项。

```
$ pg_checksums --disable -D $PGDATA
pg_checksums: syncing data directory
pg_checksums: updating control file
Checksums disabled in cluster
```

关闭校验和几乎是瞬间完成的，这是因为关闭操作不会将现有的校验和归零，而只是修改 PostgreSQL 12 的控制文件，并告诉它不要使用校验和，即使数据文件本身仍然存在校验和，也是安全的。如果再次启用校验和，那么即使已有校验和，pg_checksums 工具也会为每页生成一个校验和，并使用新的校验和重写页头。

3.2.2 COPY FROM 命令数据过滤

数据导入和导出是比较常用的功能，使用 COPY FROM 命令可以很方便地导出表的全部数据或部分数据。在 PostgreSQL12 中支持使用 where 条件扩展 COPY FROM 命令，允许数据文件在导入时进行数据的过滤。

通常，可以在源数据导出时，对查询语句使用 where 条件进行过滤。

```
COPY ( select a, b, c from tab where a ='xx' ) TO '/tmp/tab';
```

除此之外，也可以在目标数据导入时，使用 COPY FROM 命令的 where 条件。

```
COPY t_tab FROM '/tmp/tab' where a ='xx';
```

3.2.3 用户级流复制超时控制

在 PostgreSQL 12 中，可以在每个连接上设置参数 wal_sender_timeout。这对不同物理位置备库的流复制管理提供了很好的灵活性。用户能够根据备库与主库的延迟来更改参数 wal_sender_timeout 的值。例如，物理位置靠近主库的备库可以设置较小的参数 wal_sender_timeout 的值来允许快速进行问题检测和故障转移，而设置较大的参数 wal_sender_timeout 的值有助于较远位置的备库正确判断其健康状况。

参数 wal_sender_timeout 自 PostgreSQL 9.3 引入以来，只能在服务器上设置级别，并且设置之后只有发送 sighup 重新加载才能生效。

在 PostgreSQL 12 中用户能够根据物理位置的不同即时设置参数 wal_sender_timeout，并在当前会话立即生效。可以根据不同的用户设置不同的参数 wal_sender_timeout，或在参数 primary_conninfo 的 options 选项中进行设置。

```
primary_conninfo = ' options=''-c wal_sender_timeout=60000'' ...'
```

3.2.4 VACUUM 操作及 ANALYZE 操作锁跳过

在 PostgreSQL 12 中，VACUUM 操作及 ANALYZE 操作新增了一个 SKIP_LOCKED 选项可以跳过被锁住的表，使对应操作能够执行成功。PostgreSQL 12 之前的版本遇到被锁住的表时，会一直处于等待状态。

VACUUM 操作的测试过程示例如下。

在第一个会话中执行 foo 表的显式锁定操作。

```
postgres=# begin;
BEGIN
postgres=# lock table foo in exclusive mode;
LOCK TABLE
```

在第二个会话中进行 VACUUM 操作测试，不添加 SKIP_LOCKED 选项直接执行操作，会一直处于等待状态。

```
postgres=# vacuum foo;
^CCancel request sent
ERROR:  canceling statement due to user request
```

若手动取消 VACUUM 操作后，先添加 SKIP_LOCKED 选项，再执行 VACUUM 操作，则 VACUUM 操作可以执行成功，也能够看到一条警告信息。通过内置变量 SQLSTATE 查看执行结果，返回值为 00000，表示 VACUUM 操作执行成功。

```
postgres=# vacuum (SKIP_LOCKED) foo;
WARNING:  skipping vacuum of "foo" --- lock not available
VACUUM

postgres=#  \echo :SQLSTATE
00000
```

ANALYZE 操作也是如此，不添加 SKIP_LOCKED 选项直接执行 ANALYZE 操作，会一直处于等待状态。

```
postgres=# analyze foo;
^CCancel request sent
ERROR:  canceling statement due to user request
```

若手动取消 ANALYZE 操作后，先添加 SKIP_LOCKED 选项，再执行 ANALYZE 操作，则 ANALYZE 操作可以执行成功，也能够看到一条警告信息。通过内置变量 SQLSTATE 查看执行结果，返回值为 00000，表示 ANALYZE 操作执行成功。

```
postgres=# analyze (SKIP_LOCKED) foo;
WARNING:  skipping analyze of "foo" --- lock not available
ANALYZE
```

```
postgres=#  \echo :SQLSTATE
00000
```

3.2.5 表及索引清理解耦

在 MVCC 机制下，表的数据被更新或删除后，空间既不能被释放又不能立即被重用。只有执行 VACUUM 操作后，死元组的空间才能被回收重用（空间依然不会被释放给操作系统）。在执行 VACUUM 操作时，会同时对表中的索引进行清理，可以在 PostgreSQL 12 中通过表级参数进行解耦控制，如在只需要清理表，暂时不需要清理索引时，可以对表设置参数 vacuum_index_cleanup。

```
postgres=# alter table test set (vacuum_index_cleanup = false);
ALTER TABLE
```

除了可以对表设置参数，PostgreSQL 12 也允许用户在手动执行 VACUUM 操作时显式设置 INDEX_CLEANUP 选项。

```
postgres=# vacuum (VERBOSE ON, INDEX_CLEANUP OFF) test;
INFO:  vacuuming "public.test"
INFO:  "test": found 4000 removable, 0 nonremovable row versions in 18 out of 18
  pages
DETAIL:  0 dead row versions cannot be removed yet, oldest xmin: 603
There were 0 unused item identifiers.
Skipped 0 pages due to buffer pins, 0 frozen pages.
0 pages are entirely empty.
CPU: user: 0.00 s, system: 0.00 s, elapsed: 0.00 s.
VACUUM
```

这个功能可以使某些维护任务的执行速度更快。假设有一个带有很多索引的表，计划对该表重建索引，在操作之前，要先单独清理表，正常的 VACUUM 操作会清理索引。此时，就可以选择使用 INDEX_CLEANUP 选项，让其在清理过程中按照用户的需求来决定是否清理表中的索引。

3.2.6 索引在线重建

在哪些场景下需要重建索引呢，以下是几个常见的场景。
- 由于软件 BUG 或硬件原因导致的索引不可用。
- 索引中包含许多空的或近似空的页，这在 B-Tree 索引中容易发生。
- 修改了存储参数（如填充因子），确保修改生效。
- 在线创建索引时失败，遗留了一个失效的索引。

在一般情况下，很少需要重建索引。在创建索引时，PostgreSQL 提供了 CREATE INDEX CONCURRENTLY 命令，该命令允许用户在表的写负载很重时也能同时创建索引。当使用正常的 CREATE INDEX 命令创建索引时会阻塞表。因此，很难在 7×24 小时的 OLTP 中创建大型索引。

进行 DML 操作频繁的表索引可能出现较大膨胀，导致占用空间多，对性能也有一定的影响。此时，需要在数据库或某个模式中通过重建索引来修复损坏的索引或移除膨胀的索引数据。PostgreSQL 12 引入了 REINDEX CONCURRENTLY，为了解决在重建索引时不能执行查询操作的问题。REINDEX CONCURRENTLY 的大致工作流程如下。

```
create index concurrently new_index on ...;
drop index concurrently old_index;
alter index new_index rename to old_index;
```

在上述重建和移除索引的过程中，不会阻塞表中的 DML 操作，只在重命名旧索引阶段需要独占锁，以最小的锁粒度完成重建索引。

使用示例如下。

```
postgres=# REINDEX INDEX CONCURRENTLY tab_index;
REINDEX
```

注意，如果重建索引失败，那么这个索引状态将变为不可用，之后只能重建。

3.2.7 执行计划显示非默认参数

当使用 EXPLAIN 命令查看语句的执行计划时，执行计划会受许多配置参数的影响，通常会忽略有哪些非默认参数被修改过。在 PostgreSQL 12 中可以使用 settings 选项输出与执行计划相关的非默认参数，在分析 SQL 语句的性能时这个参数很有用，当 SQL 语句的性能出现问题且 SQL 语句的执行计划发生变化时可以通过此参数进行分析。

如果没有修改执行计划相关参数的默认值，那么下面的代码不会有任何变化。

```
postgres=# explain (settings on) select count(*) from pg_settings;
                           QUERY PLAN
-------------------------------------------------------------------------
 Aggregate  (cost=12.50..12.51 rows=1 width=8)
   ->  Function Scan on pg_show_all_settings a  (cost=0.00..10.00 rows=1000 width=0)
(2 rows)
```

下面修改参数 enable_seqscan，可以发现执行计划输出了修改的参数值。

```
postgres=# set enable_seqscan = false;
SET

postgres=# explain (settings on) select count(*) from pg_settings;
                           QUERY PLAN
-------------------------------------------------------------------------
 Aggregate  (cost=12.50..12.51 rows=1 width=8)
   ->  Function Scan on pg_show_all_settings a  (cost=0.00..10.00 rows=1000 width=0)
 Settings: enable_seqscan = 'off'
(3 rows)
```

下面修改更多的参数，可以看到 3 个修改的参数值。

```
postgres=# set random_page_cost = 1.1 ;
SET
```

```
postgres=# set cpu_tuple_cost = 0.02 ;
SET

postgres=# explain (settings on) select count(*) from pg_settings;
                                QUERY PLAN
-----------------------------------------------------------------------
 Aggregate  (cost=22.50..22.52 rows=1 width=8)
   ->  Function Scan on pg_show_all_settings a  (cost=0.00..20.00 rows=1000 width=0)
 Settings: cpu_tuple_cost = '0.02', enable_seqscan = 'off', random_page_cost =
    '1.1'
(3 rows)
```

auto_explain 模块也可以使用参数 log_settings。

```
postgres=# set auto_explain.log_settings = true;
SET
```

设置上面的参数后，在服务端文件中可以看到在 auto_explain 模块记录的执行计划中显示了修改的参数值。

```
Query Text: explain (settings on) select count(*) from pg_settings;
Aggregate  (cost=22.50..22.52 rows=1 width=8)
  ->  Function Scan on pg_show_all_settings a  (cost=0.00..20.00 rows=1000 width=0)
Settings: cpu_tuple_cost = '0.02', enable_seqscan = 'off', random_page_cost =
   '1.1'",,,,,,,,,,"psql"
```

3.2.8 后台操作进度报告引入

PostgreSQL 9.6 对后台操作的 VACUUM 命令提供了 pg_stat_progress_vacuum 视图，用于显示当前操作正在执行的各个阶段的信息。

PostgreSQL 12 引入了两个后台操作进度报告，可以实时观测 CREATE INDEX 命令和 CLUSTER 命令的执行状态。

在 PostgreSQL 12 之前，对大表创建索引是一个比较痛苦的过程，创建索引的过程中无法得知进度。PostgreSQL 12 新增的 pg_stat_progress_create_index 视图可以监控创建索引的进度。在创建索引时，pg_stat_progress_create_index 视图将显示当前创建索引的各个阶段的信息。

每个创建索引的进程在 pg_stat_progress_create_index 视图中对应一条记录，几个主要的字段解释如下。

- pid：创建索引的进程号。
- relid：表的对象标识符。
- index_relid：创建索引的对象标识符。
- phase：创建索引的当前处理阶段。
- current_locker_pid：阻塞索引创建的进程号。
- blocks_total：当前处理阶段需要处理的数据块。
- lockers_done：当前处理阶段已完成的数据块。
- tuples_total：当前处理阶段需要处理的记录数。

- tuples_done：当前处理阶段已完成的记录数。
- partitions_total：在分区表上创建索引时，当前处理阶段需要处理的总分区数。
- partitions_done：在分区表上创建索引时，当前处理阶段已处理的总分区数。

由于创建索引的时间主要花费在 building index: scanning table 阶段和 building index: loading tuples in tree 阶段，因此根据 blocks_total 字段、blocks_done 字段、tuples_total 字段会比较容易判断创建索引的进度。

CLUSTER 命令的作用是依据索引对表数据排序。使用 CLUSTER 命令和 VACUUM FULL 命令都可以物理移动数据。在运行 CLUSTER 命令或 VACUUM FULL 命令时，pg_stat_progress_cluster 视图用于显示当前操作正在执行的各个阶段的信息。

3.2.9 备库升主库开放 SQL 接口

激活流复制备库是非常重要的运维管理操作，在 PostgreSQL 12 之前的版本中有两种方式可以激活流复制备库。

- pg_ctl 命令方式：在备库主机上执行 pg_ctl promote 命令。
- 触发器文件方式：先配置参数 trigger_file，再在备库主机上创建触发器文件。

这两种方式都必须登录备库主机进行操作。PostgreSQL 12 中新增了 pg_promote 函数，允许在 SQL 接口激活备库，将备库升为主库。通过 SQL 接口去执行的这种激活方式显然更灵活。

赋予普通用户 pg_promote 函数的权限，而不需要超级用户权限。

```
CREATE ROLE promote_role LOGIN;

GRANT EXECUTE ON FUNCTION pg_promote TO promote_role;
```

使用普通用户通过 SQL 语句执行备库升主库操作。

```
select pg_promote();
```

3.2.10 PSQL 工具帮助链接添加

PSQL 工具中的帮助命令在 PostgreSQL 12 之前的版本中并没有关联官方文档的直接链接，PostgreSQL 12 在每个帮助命令底部都添加了对应的官方文档链接。在 show 命令底部添加对应的官方文档链接的示例如下。

```
postgres=# \h show
Command:     SHOW
Description: show the value of a run-time parameter
Syntax:
SHOW name
SHOW ALL
URL: https://www.postgresql.org/docs/12/sql-show.html
```

上面的 URL 非常有用，可以避免用户进行一些不必要的查找。

3.3 PostgreSQL 12 的开发易用性提升

PostgreSQL 12 的开发易用性提升主要体现在声明式分区增强、运算存储列使用、绑定变量窥探引入、SQL/JSON path 引入、枚举数据类型增强。

3.3.1 声明式分区增强

在使用声明式分区时有许多使用限制，其中比较重要的限制之一是普通表不允许外键引用分区表。PostgreSQL 12 中的外键和分区表完全兼容，可以和普通表一样使用外键约束。

下面是一个简单的示例。

```
CREATE TABLE items (
    item_id integer PRIMARY KEY,
    description text NOT NULL
) PARTITION BY hash (item_id);

CREATE TABLE items_0 PARTITION OF items FOR VALUES WITH (modulus 3, remainder 0);

CREATE TABLE items_1 PARTITION OF items FOR VALUES WITH (modulus 3, remainder 1);

CREATE TABLE items_2 PARTITION OF items FOR VALUES WITH (modulus 3, remainder 2);

CREATE TABLE warehouses (warehouse_id integer primary key, location text not null);

CREATE TABLE stock (
    item_id integer not null REFERENCES items,
    warehouse_id integer not null REFERENCES warehouses,
    amount int not null
) partition by hash (warehouse_id);

CREATE TABLE stock_0 PARTITION OF stock FOR VALUES WITH (modulus 3, remainder 0);

CREATE TABLE stock_1 PARTITION OF stock FOR VALUES WITH (modulus 3, remainder 1);

CREATE TABLE stock_2 PARTITION OF stock FOR VALUES WITH (modulus 3, remainder 2);
```

stock 表有两个外键，一个指向未分区的 warehouses 表，一个指向分区的 items 表，直接在创建表的语句中指定 REFERENCES 即可。

PostgreSQL 12 之前的版本只允许将简单常量用作分区边界，而 PostgreSQL 12 允许分区边界是任意的表达式。

```
postgres12=# CREATE TABLE teachers_30s PARTITION OF
teachers FOR VALUES FROM (10*3) TO (10*3+10);
CREATE TABLE
CREATE TABLE t1 PARTITION OF t FOR VALUES FROM (10*3) TO (10*3+10);
```

另外，PostgreSQL 12 对普通表连接分区的操作降低了锁粒度，PostgreSQL 12 之前的版本在父表和被连接分区中都要添加 Access Exclusive 锁。PostgreSQL 12 只会在被连接分区和默认分区（默认分区可选）中添加 Access Exclusive 锁，父表只需要添加 Share Update Exclusive 锁。

对于一维分区表，使用 PostgreSQL 12 提供的元命令足够查看分区的完整信息；但对于多维分区表，使用元命令无法查看详细的分区信息。使用 PostgreSQL 12 提供的如下分区函数，可以很方便地查看分区信息。

- pg_partition_tree（regclass）：返回分区表的详细信息，如分区名称、上一级分区名称、是否叶子结点、层级。
- pg_partition_ancestors（regclass）：返回上层分区名称，包括本层分区名称。
- pg_partition_root（regclass）：返回顶层父表名称。

3.3.2 运算存储列使用

在某些情况下，某列的值需要通过其他列进行运算动态产生，通常需要使用触发器来实现。然而，编写触发器需要额外的编码工作，PostgreSQL 12 提供了更好的解决方案。

假设以千米和海里为单位存储数据。1 海里约等于 1.852 千米，为了确保海里能自动生成，可以使用下面的代码。

```
postgres=# CREATE TABLE t_measurement (
t    timestamp,
km   numeric,
nm   numeric GENERATED ALWAYS AS (km * 1.852) STORED
);
CREATE TABLE
```

使用 GENERATED ALWAYS AS 语句来计算列的值。

下面的代码中插入的为 100 千米，通过自动计算返回的为 185.200 海里，符合预期。

```
postgres=# insert into t_measurement (t,km) values(now(),100) returning *;
             t              | km  |   nm
----------------------------+-----+---------
 2020-08-25 14:29:55.548074 | 100 | 185.200
(1 row)
INSERT 0 1
```

通过 GENERATED ALWAYS 语句声明的列，插入及修改语句不能显式赋值，以确保列的值始终按照系统设置而不会被修改。

如果对千米数进行更新，那么海里数也会自动更新。下面的代码也是符合预期的。

```
postgres=# update t_measurement set km=200 where km=100 returning * ;
             t              | km  |   nm
----------------------------+-----+---------
 2020-08-25 14:29:55.548074 | 200 | 370.400
(1 row)
UPDATE 1
```

运算存储列有如下限制。

- 运算存储列只能引用本表的非运算存储列字段，不可以引用其他表的字段。
- 运算存储列使用的表达式和操作符必须为 IMMUTABLE 属性。
- 运算存储列的字段为只读属性，不支持 INSRET 操作和 UPDATE 操作。

- 运算存储列不能作为分区键使用。

3.3.3 绑定变量窥探引入

在介绍绑定变量之前，下面了解一下硬解析与软解析。硬解析是指 SQL 语句经历完整的语法及语义分析、SQL 语句解析、执行计划生成、SQL 语句执行等阶段。而软解析则是指复用已有的 SQL 语句执行计划，直接执行。使用软解析的优点是可以避免硬解析与之相关的额外开销，缺点是绑定变量在被使用时，查询优化器会忽略其具体值，其预估的准确性在表存在数据倾斜时会提供错误的执行计划。

Oracle 通过绑定变量窥探（Bind Peeking）对硬解析进行实际窥探来提高 SQL 语句执行计划的准确性。PostgreSQL 中也有类似的概念。执行计划有以下两种策略。
- custom plan：重新生成执行计划，相当于 Oracle 中的硬解析。
- generic plan：生成可重用的通用执行计划，相当于 Oracle 中的软解析。

PostgreSQL 12 中新增了参数 plan_cache_mode，可以设置优化器策略。
- auto：默认的优化器策略，前 5 次使用硬解析产生 custom plan，第 6 次开始考虑使用 generic plan。
- force_custom_plan：强制优化器策略固定为 custom plan。
- force_generic_plan：强制优化器策略固定为 generic plan。

下面通过示例进行演示。

```
create table t (id int primary key);

insert into t select * from generate_series(1,10000) ;
```

在创建测试表并插入数据之后，参数 plan_cache_mode 使用默认的 auto 进行测试。

```
postgres=# show plan_cache_mode ;
 plan_cache_mode
-----------------
 auto
(1 row)
```

使用 PostgreSQL 12 的 PREPARE 语句，可以避免出现反复解析。

```
postgres=# prepare s(int) as select * from t where id = $1;
PREPARE
```

使用 EXECUTE 语句进行测试。

```
postgres=# explain execute s(100);
                    QUERY PLAN
-----------------------------------------------------------------
 Index Only Scan using t_pkey on t  (cost=0.29..8.30 rows=1 width=4)
   Index Cond: (id = 100)
(2 rows)
```

EXECUTE 语句第 6 次执行的结果如下。

```
postgres=# explain execute s(100);
                    QUERY PLAN
-----------------------------------------------------------------
 Index Only Scan using t_pkey on t  (cost=0.29..8.30 rows=1 width=4)
```

```
    Index Cond: (id = $1)
(2 rows)
```

观察执行计划中的 Index Cond 所在行，对比可以发现，前 5 次执行计划中的条件使用常量 100，也就是采用 custom plan；第 6 次执行计划中的条件使用变量$1，也就是采用 generic plan。

在默认的优化器策略下，如果查询列数据分布均匀无倾斜，那么第 6 次将生成稳定的执行计划，通过绑定变量窥探可以减少硬解析的次数。如果在查询列数据分布不均的情况下生成通用执行计划，那么在绑定变量的值发生倾斜时会出现使用错误执行计划的情况，这时可以把参数 plan_cache_mode 的值设置成 force_custom_plan，避免因执行计划缓存而出错。

3.3.4 SQL/JSON path 引入

与开发人员高度相关的一个特性是，PostgreSQL 12 提供了大量的新函数来更快捷地解析 JSON 数据类型，这也极大地提高了 PostgreSQL 12 处理 NoSQL 风格负载的能力。SQL/JSON path 的核心是定义函数表达式，实现方式是使用 SQL/JSON path 数据类型，SQL/JSON path 以二进制形式展现函数表达式。

下面是 SQL/JSON path 函数表达式使用 JavaScript 的一些语法。
- 点号表示引用 JSON 数据中的元素。
- 方括号表示引用数组元素。
- JSON 数据中的数组元素下标从 0 开始。

下面是 SQL/JSON path 函数表达式的变量。
- $表示要查询的 JSON 文本的变量。
- $varname 表示指定变量。
- @ 表示在 filter 表达式中当前路径元素的变量。

为了方便演示，下面创建测试表并插入一条 JSON 数据。

```
CREATE TABLE T_JSONPATH (a jsonb);

INSERT INTO T_JSONPATH (a) VALUES ('
{ "gpsname": "postgres",
  "track" :
  {
   "segments" : [
     { "location":   [ 49.773, 15.2104 ],
       "start time": "2020-05-11 10:05:14",
       "HR": 73
     },
     { "location":   [ 49.776, 15.4125 ],
       "start time": "2020-06-21 10:39:21",
       "HR": 130
     } ]
  }
}');
```

使用 jsonb_pretty 函数进行 JSON 数据的格式化输出。

```
postgres=# SELECT jsonb_pretty(a) FROM T_JSONPATH;
                jsonb_pretty
-----------------------------------------------------
 {                                                  +
     "track": {                                     +
         "segments": [                              +
             {                                      +
                 "HR": 73,                          +
                 "location": [                      +
                     49.773,                        +
                     15.2104                        +
                 ],                                 +
                 "start time": "2020-05-11 10:05:14"+
             },                                     +
             {                                      +
                 "HR": 130,                         +
                 "location": [                      +
                     49.776,                        +
                     15.4125                        +
                 ],                                 +
                 "start time": "2020-06-21 10:39:21"+
             }                                      +
         ]                                          +
     },                                             +
     "gpsname": "postgres"                          +
 }
(1 row)
```

PostgreSQL 12 之前的版本可以通过操作符查询 JSON 数据中的元素。

```
postgres=# SELECT a ->> 'gpsname' FROM T_JSONPATH;
 ?column?
----------
 postgres
(1 row)
```

PostgreSQL 12 可以使用 SQL/JSON path 函数表达式进行查询。

```
postgres=# SELECT jsonb_path_query(a,'$.gpsname') FROM T_JSONPATH;
 jsonb_path_query
------------------
 "postgres"
(1 row)
```

以上代码中使用了 jsonb_path_query 函数，这个函数是 SQL/JSON path 的常用函数。

如果 JSON 数据涉及较多层级，那么 SQL/JSON path 函数表达式就更加易用。例如，查询 T_JSONPATH 表的 track.segments 下一层级的元素。

```
postgres=# SELECT jsonb_path_query(a,'$.track.segments[1].HR') FROM T_JSONPATH;
 jsonb_path_query
------------------
 130
(1 row)
```

除此之外，还可以使用 jsonb_path_exists 函数判断是否存在指定 JSON 数据的路径。

```
postgres=# SELECT jsonb_path_exists(a,'$.track.segments.HR') FROM T_JSONPATH;
 jsonb_path_exists
-------------------
 t
(1 row)
postgres=# SELECT jsonb_path_exists(a,'$.track.segments.ab') FROM T_JSONPATH;
 jsonb_path_exists
-------------------
 f
(1 row)
```

3.3.5 枚举数据类型增强

PostgreSQL 12 对枚举数据类型进行了增强，可以在单个事务内修改枚举数据类型，但是需要注意一个限制，即在事务中访问新建的枚举数据类型会提示错误。

下面创建货币枚举数据类型 currency，在事务中添加新货币。

```
postgres=# create type currency as enum ('USD','EUR','GBP');
CREATE TYPE
postgres=# begin ;
BEGIN
postgres=# alter type currency add value 'CHF' after 'EUR';
ALTER TYPE
postgres=# select 'USD'::currency;
 currency
----------
 USD
(1 row)
```

添加的货币 CHF 必须在提交之后使用，否则会报错。

```
postgres=# select 'CHF'::currency;
ERROR:  unsafe use of new value "CHF" of enum type currency
LINE 1: select 'CHF'::currency;
               ^
HINT:  New enum values must be committed before they can be used.
```

3.4 PostgreSQL 12 的系统层变化

PostgreSQL 12 的系统层变化主要体现在表存储引擎开放、恢复相关配置优化、系统元数据引入、配置参数引入、流复制连接数优化、DOS 攻击预防、SSL 协议可控，以及附加模块变化。

3.4.1 表存储引擎开放

PostgreSQL 全球社区一直把注意力和开发精力集中在通用的堆存储引擎上。虽然通用的堆存储引擎在许多情况下运行性能良好，但在某些情况下仍然需要不同的存储方法。PostgreSQL 12 的系统层比较重要的一个变化是引入了 Pluggable Table Storage Interface，为后续支持多种存储引擎奠定了基础，如 zheap、memory、columnar-oriented 等存储引擎。

存储引擎的创建可以通过 CREATE EXTENSION 命令创建外部扩展实现，之后在创建表时指定新建的存储引擎即可，可以在会话层设置存储引擎，之后创建的表默认使用已设置的存储引擎。

在 PostgreSQL 12 中，通过参数 default_table_access_method 可以查看表的默认存储引擎。

```
postgres=# show default_table_access_method;
 default_table_access_method
-----------------------------
 heap
(1 row)
```

3.4.2 恢复相关配置优化

在 PostgreSQL 12 之前的版本中，如果数据目录中存在 recovery.conf 文件，那么当实例启动时将进入恢复模式（recovery 模式或 standby 模式），该文件包含了用于恢复配置的所有参数。

- standby_mode：确定是正常的归档恢复还是 standby 模式。
- restore_command：恢复已归档的 WAL 文件。
- recovery_target*：确定要恢复到的点。
- primary_conninfo：确定如何连接到流复制主服务器上。

长期以来，独立的 recovery.conf 文件配置一直被认为是一个设计缺陷，这是因为参数分布在多个不同的文件中是不合理的，另外也不能使用 ALTER SYSTEM 命令对参数进行修改。

从 PostgreSQL 12 开始，针对此问题进行了改进，即把 recovery.conf 文件中的参数合并到 postgresql.conf 文件中，应注意在非恢复模式下这些参数将被忽略。

那么如何让 PostgreSQL 知道它处于恢复模式呢？

在 PostgreSQL 12 之前的版本中存在 recovery.conf 文件，即触发恢复模式。

从 PostgreSQL 12 开始不存在该文件，由下面两个新文件代替。

- recovery.signal：告诉 PostgreSQL 进入正常的归档恢复。
- standby.signal：告诉 PostgreSQL 进入 standby 模式。

如果这两个文件都存在，那么优先使用 standby.signal 文件。

3.4.3 系统元数据引入

PostgreSQL 12 新引入了 5 个系统元数据，如表 3-1 所示。

表 3-1 PostgreSQL 12 新引入的系统元数据

元数据名称	描述
pg_stat_progress_cluster	显示 CLUSTER 命令执行的状态信息
pg_stat_progress_create_index	显示 CREATE INDEX 命令执行的状态信息
pg_statistic_ext_data	显示规划器统计信息
pg_stats_ext	显示 pg_statistic_ext_data 视图中的公共可读信息,仅显示有关当前用户可读的表及列信息
pg_stat_gssapi	显示 GSSAPI 的使用情况

3.4.4 配置参数引入

PostgreSQL 12 新引入了 11 个配置参数,如表 3-2 所示。

表 3-2 PostgreSQL 12 新引入的配置参数

参数名称	描述
default_table_access_method	在指定 CREATE TABLE 或 CREATE MATERIALIZED VIEWS 不带访问方法时,TABLE 或 MATERIALIZED VIEWS 的访问方法,以及使用 select…into 语句时用到的访问方法
log_transaction_sample_rate	设置除其他原因记录的语句外,所有记录的语句的事务百分比
plan_cache_mode	优化器策略,可以设置值为 auto、force_custom_plan、force_generic_plan
promote_trigger_file	指定 trigger file,该文件用于 Standby Server 结束 recovery 状态进行角色提升
shared_memory_type	指定 Shared Memory 的实现方式。mmap 是指使用 mmap 分配的 Anonymous Shared Memory;sysv 是指通过 shmget 分配的 System V Shared Memory;windows 是指 Windows Shared Memory
ssl_library	PostgreSQL 12 在编译时的 SSL Library
ssl_max_protocol_version	设置使用的 SSL/TLS 协议的最大版本,这对测试或有些组件使用新版本协议出现的问题是有帮助的
ssl_min_protocol_version	设置使用的 SSL/TLS 协议的最小版本,这对测试或有些组件使用新版本协议出现的问题是有帮助的
tcp_user_timeout	指定在强制关闭 TCP 连接之前,传输的数据可能保持未确认状态的时间
wal_init_zero	当值为 on 时,会导致新 WAL 文件以零填充,在某些文件系统中,这确保了在需要写入 WAL Records 之前的空间已被分配
wal_recycle	当值为 on 时,会导致 WAL 文件被循环使用(通过重命名 WAL 文件的方式),避免了创建新 WAL 文件的需要

3.4.5　流复制连接数优化

在 PostgreSQL 12 中用于控制总连接数的参数为 max_connections，应用程序、pg_basebackup 工具基础全备、流复制连接都会占用参数 max_connections 设置的连接数。如果应用程序使用的连接数膨胀，占满了连接，那么会导致数据库备份、流复制和运维操作因连接数耗尽而无法进行。

与连接数相关的 3 个参数说明如下。
- max_connections：用于设置数据库实例允许的最大并发连接数。
- max_wal_senders：用于设置通过 pg_basebackup 工具备份或流复制保持备库和主库同步占用主库的最大并发连接数。
- superuser_reserved_connections：用于设置给超级用户预留的连接数。

后两个参数设置的连接数是第一个参数的子集，PostgreSQL 12 之前的版本中的 3 个参数的配置关系如下。

```
max_connections > ( max_wal_senders + superuser_reserved_connections )
```

虽然参数 superuser_reserved_connections 可以给超级用户预留连接数，但是使用参数 superuser_reserved_connections 依然不是一个妥当的方式。例如，如果流复制用户不是超级用户，那么当主库连接数占满时，将无法保持备库和主库同步。

PostgreSQL 12 对流复制使用参数 max_wal_senders 占用的连接数从参数 max_connections 中进行了剥离，能够将应用程序占用的连接数和数据库备份、流复制占用的连接数分开。

3.4.6　DOS 攻击预防

在 PostgreSQL 11 中普通用户可能会锁住数据库连接，在进行数据库登录时受到 DOS 攻击。

例如，在会话一中，执行如下语句。

```
BEGIN;
SELECT count(*) FROM pg_stat_activity;
```

此时，会获取 pg_authid 表的共享锁。在会话二中，普通用户对 pg_authid 表执行 VACUUM FULL 操作。

```
VACUUM FULL pg_authid;
```

虽然普通用户不是 pg_authid 表的宿主，上面的操作会执行失败，但是会话二中的操作会尝试去获取该表的锁，受会话一中手动事务控制的影响，会话一在提交之前将一直处于阻塞状态。要执行 VACUUM FULL 操作需要获取 pg_authid 表的独占锁，这将阻止任何读写请求，而由于 pg_authid 表是系统登录认证过程的入口，因此任何新连接请求在会话一中的事务提交之前都不能成功。

除了执行 VACUUM FULL 操作，执行 TRUNCATE 操作也会产生类似问题，PostgreSQL 12 对这两个问题进行了修复，可以预防受到 DOS 攻击。

3.4.7 SSL 协议可控

PostgreSQL 12 对传输加密 SSL 协议增加了控制强制执行 SSL 连接时使用的协议版本，下面是两个新 GUC 参数。

- ssl_min_protocol_version，控制使用 SSL 协议的最小版本。
- ssl_max_protocol_version，控制使用 SSL 协议的最大版本。

在特定协议版本上进行开发或功能测试时，明确需要强制执行给定的版本是有意义的。

3.4.8 附加模块变化

PostgreSQL 12 中附加模块的变化具体如下。

- pg_stat_statements 模块的 pg_stat_statements_reset 函数允许重置特定数据库、用户及查询的统计信息。
- postgres_fdw 模块的重大变化是支持远程 session 级别的 GUC 参数，在创建服务器时可以使用-c 选项指定参数，多个参数之间使用空格分隔。

在下面的代码中，创建服务器时指定远程连接使用的参数 work_mem 的值为 64MB。

```
create server remote_server foreign data wrapper postgres_fdw
options (host '127.0.0.1',
        port '5432',
        dbname 'mydb',
        options '-c work_mem=64MB'
);
```

- amcheck 模块的 bt_index_parent_check 函数新增从树的根部检查每个索引元组的功能。

3.5 本章小结

PostgreSQL 12 的性能有大量的提升，包括 WITH 语句的物化开关及默认行为控制，绑定变量窥探引入，当发生数据倾斜时可以关闭软解析，强制使用硬解析。此外，与开发人员高度相关的一个特性是，PostgreSQL 12 提供了大量 SQL/JSON path 函数表达式来更快捷地解析 JSON 数据类型，这也极大地提高了 PostgreSQL 12 处理 NoSQL 风格负载的能力。在流复制方面，PostgreSQL 12 优化了恢复相关配置，将配置参数统一合并到主配置文件中，用户能够根据备库或级联备库的延迟来设置不同的参数 wal_sender_timeout。同时，将备库升为主库也新增了更加灵活的 SQL 接口方式。在运维管理优化方面，对多个索引的单表增加了索引与表的清理解耦功能，这可以让维护任务更快执行，后台操作进度报告不断丰富。

在 PostgreSQL 12 中能在线操作损坏索引的重建，可以解决重建期间不能执行查询操作的问题。在系统层的安全性方面，PostgreSQL 12 对执行 VACUUM FULL 操作或 TRUNCATE 操作可能产生的数据库被锁问题进行了修复，可以预防 DOS 攻击。

第 4 章

PostgreSQL 13 新特性

4.1 PostgreSQL 13 的主要性能提升

与 PostgreSQL 13 之前的版本相比，PostgreSQL 13 的性能有了一定的提升，主要体现在索引去重、增量排序、库级索引并发重建、HASH 聚合可溢出到磁盘、索引并行清理、PL/pgSQL 提速、Windows 连接优化。

4.1.1 索引去重

迄今为止，B-Tree 索引是在 PostgreSQL 中使用的一种非常重要的索引，大多数索引都是基于 B-Tree 索引创建的。B-Tree 索引元组的存储结构是(value,ctid)，当表的数据列含有大量重复数据时，在 PostgreSQL 13 之前的版本中，每个索引都是在索引文件中独立存储的，也就是说，索引文件会存储所有重复的数据列，索引文件中会存在大量重复数据。PostgreSQL 13 在 B-Tree 索引的存储层面引入了一个 Deduplication 去重技术，能将重复的索引合并，大幅度地减少重复索引的数量。

下面通过 PostgreSQL 12 和 PostgreSQL 13 的示例演示索引去重的过程。

在 PostgreSQL 12 中执行以下代码。

```
create table duplindex_test(a int ,b int);
create index idx_norm on duplindex_test using btree(b);
insert into duplindex_test select 3,5 from generate_series(1,100000);
```

查看索引的大小。

```
postgres=# \di+ idx_norm
```

```
List of relations
-[ RECORD 1 ]---------------
Schema      | public
Name        | idx_norm
Type        | index
Owner       | postgres
Table       | duplindex_test
Size        | 2072 kB
Description |
```

从以上代码中可以看出，在 PostgreSQL 12 中，idx_norm 索引的大小为 2072KB。

在 PostgreSQL 13 中执行以下代码。

```
create table duplindex_test(a int ,b int) ;
create index idx_norm on duplindex_test using btree(b) WITH (deduplicate_items = off) ;
create index idx_dupl on duplindex_test using btree(b) WITH (deduplicate_items = on) ;
insert into duplindex_test select 3,5 from generate_series(1,100000) ;
```

查看索引的大小。

```
postgres=# \di+ idx_norm
List of relations
-[ RECORD 1 ]---------------
Schema      | public
Name        | idx_norm
Type        | index
Owner       | postgres
Table       | duplindex_test
Size        | 2072 kB
Description |
```

从以上代码中可以看出，未使用 Deduplication 去重技术（deduplicate_items = off）的 idx_norm 索引的大小与 PostgreSQL 13 之前的版本一样。

```
postgres=# \di+ idx_dupl
List of relations
-[ RECORD 1 ]---------------
Schema      | public
Name        | idx_dupl
Type        | index
Owner       | postgres
Table       | duplindex_test
Persistence | permanent
Size        | 648 kB
Description |
```

从以上代码中可以看出，使用 Deduplication 去重技术（deduplicate_items = on）的 idx_dupl 索引的大小为 648KB。

通过上面的示例可以看出，PostgreSQL 13 在使用 Deduplication 去重技术（deduplicate_items = on）创建索引时，索引的大小将缩减为 PostgreSQL 12 的 1/3。

由此可知，Deduplication 去重技术具有以下优点。

- 减小存储空间：重复的索引被合并，减小索引的存储空间。

- 减少维护开销：加快重建索引的速度，减少 VACUUM 索引的开销。
- 提高查询效率：越小的索引的查询效率越高，可以提高系统 I/O 性能。

注意，要使用 pg_upgrade 工具升级 PostgreSQL 13 之前的版本到 PostgreSQL 13，需要通过 REINDEX 命令重建索引。

4.1.2 增量排序

PostgreSQL 13 围绕索引和排序功能进行了一些改变来提高数据库的性能，索引的改进主要是索引去重，排序的改进包括多字段的增量排序（Incremental Sort）。通过增量排序可以加快数据的排序速度，其原理是借助有序的索引来实现。如果一个表中有 a 和 b 两个字段，其中在 a 字段上建立了索引，由于索引是有序的，a 字段已经预先进行了排序，因此要同时对 a 字段和 b 字段排序只需要在 a 字段的基础上对 b 字段排序即可。

下面通过示例演示增量排序的过程。

创建测试表，插入测试数据之后对 a 字段创建索引。

```
CREATE TABLE incresort_test(a int,b int);

INSERT INTO incresort_test(a,b) SELECT n,round(random()*100000000) FROM
   generate_series(1,1000000) n;

CREATE INDEX idx_incresort_test ON incresort_test USING btree(a);
```

执行下面的代码，结果集按 a 字段和 b 字段进行排序。

```
postgres=# EXPLAIN  SELECT * FROM incresort_test ORDER BY a,b;
                             QUERY PLAN
-------------------------------------------------------------------------
 Incremental Sort  (cost=0.47..75408.43 rows=1000000 width=8)
   Sort Key: a, b
   Presorted Key: a
   ->  Index Scan using idx_incresort_test on incresort_test  (cost=0.42..30408.42
   rows=1000000 width=8)
(4 rows)
```

重点关注上面的执行计划中的 Incremental Sort 和 Presorted Key:a 所在行可以看出，使用 a 字段的 idx_incresort_test 索引对 b 字段进行了增量排序。

```
postgres=# EXPLAIN ANALYZE SELECT * FROM incresort_test ORDER BY a,b;
                             QUERY PLAN
-------------------------------------------------------------------------
 Incremental Sort  (cost=0.47..75408.43 rows=1000000 width=8) (actual
   time=0.052..608.216 rows=1000000 loops=1)
   Sort Key: a, b
   Presorted Key: a
   Full-sort Groups: 31250  Sort Method: quicksort  Average Memory: 26kB  Peak
   Memory: 26kB
   ->  Index Scan using idx_incresort_test on incresort_test  (cost=0.42..30408.42
   rows=1000000 width=8) (actual time=0.025..280.557 rows=1000000 loops=1)
 Planning Time: 0.072 ms
 Execution Time: 691.267 ms
```

(7 rows)

观察真实执行计划可以发现,执行时间为 691.267ms。

增量排序通过参数 enable_incremental_sort 控制,默认值为 on,设置值为 off 之后,观察上面的示例在 PostgreSQL 13 之前版本中的行为。

```
postgres=# SET enable_incremental_sort TO off;
SET
```

关闭增量排序之后,再次观察执行计划。

```
postgres=# EXPLAIN ANALYZE SELECT * FROM incresort_test ORDER BY a,b;
                         QUERY PLAN
-------------------------------------------------------------------------
 Sort  (cost=127757.34..130257.34 rows=1000000 width=8) (actual
   time=663.148..945.270 rows=1000000 loops=1)
   Sort Key: a, b
   Sort Method: external merge  Disk: 17696kB
   ->  Seq Scan on incresort_test  (cost=0.00..14425.00 rows=1000000 width=8)
   (actual time=0.011..167.902 rows=1000000 loops=1)
 Planning Time: 0.101 ms
 Execution Time: 1039.032 ms
(6 rows)
```

从以上代码中可以看出,将返回旧的排序机制,执行时间为 1039.032ms。对比可知,使用增量排序更节省执行时间。不仅如此,旧的排序机制必须对整个数据集进行排序,对内存的要求较大。换句话说,使用增量排序不仅可以使性能得到提升,而且可以大大减少内存消耗。

4.1.3 库级索引并发重建

REINDEXDB 命令用于重建数据库中的索引,既可以对 Schema 级别索引进行重建,又可以对 Database 级别索引进行重建。PostgreSQL 13 对 REINDEXDB 命令添加了 -j 选项,可以使用多个并发连接进行索引重建,从而加快重建索引的速度。

在下面的示例中,示例一使用了 -j 选项,并对 mydb 数据库使用了 8 个并发连接进行索引重建,示例二没有使用 -j 选项,只使用了一个连接进行操作。

示例一:
```
$ time reindexdb -j 8 mydb
real 0m8.521s
user 0m0.012s
sys 0m0.008s
```

示例二:
```
$ time reindexdb mydb
real 0m47.137s
user 0m0.001s
sys 0m0.008s
```

通过上面示例的整体执行时间对比可以看出,使用 8 个并发连接可以加速重建索引的过程。不过只有在索引数量足够的情况下使用 -j 选项才会有效,并且数据库中的表越大,使用该选项的效果越明显。注意,使用 -j 选项会给数据库主机带来压力,用户需要根据服务端配置设置并发数,并在业务低谷期操作。

4.1.4　HASH聚合可溢出到磁盘

在运行简单的 GROUP BY 子句时,执行器有两种基本的实现方式:GroupAggregate 与 HashAggregate。

GroupAggregate 通常用于分组数量特别大的场景。例如,要对数十亿人按电话号码进行分组统计,由于分组数量太大,因此 PostgreSQL 无法在内存中进行操作,此时 GroupAggregate 会被执行。HashAggregate 通常用于分组数量较小的场景。例如,要对数十亿人按性别进行分组统计,由于分组数量较小,因此 PostgreSQL 可以在内存中进行操作,此时 HashAggregate 会被执行。

如果 PostgreSQL 对分组数量预估错误,那么执行 HashAggregate 会造成内存占用过大,甚至引起 OOM。目前,PostgreSQL 13 允许 HashAggregate 溢出到磁盘,当 HASH 操作占用的内存超过 work_mem * hash_mem_multiplier 的值时,剩余的内存占用会被溢出到磁盘。

下面通过示例演示 HASH 聚合溢出到磁盘的过程。

创建测试表并插入 10 万条测试数据。

```
create table hashagg_test (a int, b varchar(30)) ;

insert into hashagg_test select a, 'this is a test'||a from
    generate_series(1,100000) a;
```

对 a 字段分组统计条数。

```
postgres=# explain analyze select count(*) from hashagg_test group by a;
                                    QUERY PLAN
-------------------------------------------------------------------------
 HashAggregate  (cost=1344.07..1346.07 rows=200 width=12) (actual
   time=104.347..181.805 rows=100000 loops=1)
   Group Key: a
   Batches: 5  Memory Usage: 4161kB  Disk Usage: 1760kB
   ->  Seq Scan on hashagg_test  (cost=0.00..1108.38 rows=47138 width=4) (actual
   time=0.065..30.053 rows=100000 loops=1)
 Planning Time: 0.783 ms
 Execution Time: 195.150 ms
(6 rows)


Time: 312.893 ms
```

从真实执行计划中可以看出,超过参数 work_mem 设置的默认值 4MB 的 1760KB 溢出到了磁盘。

4.1.5　索引并行清理

VACUUM 操作是 PostgreSQL 中一个常用的功能。从 PostgreSQL 13 开始,可以对单个表对象的索引进行并行清理,表中的每个索引都可以分配一个工作清理进程。它的使用方式如下。
```
VACUUM (parallel 2) some_table;
```
  VACUUM 操作的并行度可以简单地使用核数来进行设置，充分利用 CPU 的核心，但最大不能超过参数 max_parallel_maintenance_workers 的值。

  下面通过示例演示索引并行清理的过程。

  首先创建有 3 个字段的测试表，对 3 个字段分别创建索引，其次插入测试数据，并进行 UPDATE 操作，产生需要清理的死元组。

```
create table vacparal_test (a int,b varchar,c varchar);
create index idx_vacparal_a on vacparal_test (a);
create index idx_vacparal_b on vacparal_test (b);
create index idx_vacparal_c on vacparal_test (c);
insert into vacparal_test select id,'this is a vacuum parallel test','this is the
    last column' from generate_series(1,1000000) id;
update vacparal_test set b = 'vacuum parallel test' where a < 100000 ;
```

  3 个字段的索引如下。

```
postgres=# \di+ idx_vacparal_a
                         List of relations
 Schema |      Name      | Type  |  Owner   |     Table     | Size  | Description
--------+----------------+-------+----------+---------------+-------+-------------
 public | idx_vacparal_a | index | postgres | vacparal_test | 22 MB |
(1 row)

postgres=# \di+ idx_vacparal_b
                         List of relations
 Schema |      Name      | Type  |  Owner   |     Table     | Size  | Description
--------+----------------+-------+----------+---------------+-------+-------------
 public | idx_vacparal_b | index | postgres | vacparal_test | 45 MB |
(1 row)

postgres=# \di+ idx_vacparal_c
                         List of relations
 Schema |      Name      | Type  |  Owner   |     Table     | Size  | Description
--------+----------------+-------+----------+---------------+-------+-------------
 public | idx_vacparal_c | index | postgres | vacparal_test | 37 MB |
(1 row)
```

  下面并行清理索引。

```
postgres=# vacuum (verbose on,parallel 2) vacparal_test ;
INFO:  vacuuming "public.vacparal_test"
INFO:  launched 2 parallel vacuum workers for index cleanup (planned: 2)
INFO:  index "idx_vacparal_b" now contains 4845 row versions in 888 pages
DETAIL:  0 index row versions were removed.
79 index pages have been deleted, 79 are currently reusable.
CPU: user: 0.00 s, system: 0.00 s, elapsed: 0.00 s.
INFO:  index "idx_vacparal_c" now contains 4763 row versions in 879 pages
DETAIL:  0 index row versions were removed.
79 index pages have been deleted, 79 are currently reusable.
CPU: user: 0.00 s, system: 0.00 s, elapsed: 0.00 s.
```

```
INFO:  "vacparal_test": found 0 removable, 54 nonremovable row versions in 1 out of
    12395 pages
DETAIL:  0 dead row versions cannot be removed yet, oldest xmin: 1385
There were 0 unused item identifiers.
Skipped 0 pages due to buffer pins, 1136 frozen pages.
0 pages are entirely empty.
CPU: user: 0.02 s, system: 0.08 s, elapsed: 0.12 s.
VACUUM
Time: 139.608 ms
```

从上面的清理过程可以看出，VACUUM 操作根据并行度设置启动两个并行工作进程来清理索引，如果不启动并行工作进程，那么耗时将增加一倍。因此，启动并行工作进程会提高清理效率。

4.1.6　PL/pgSQL 提速

PL/pgSQL 在函数或存储过程中使用较多，PostgreSQL 13 帮助开发人员加速了 PL/pgSQL 对循环代码的处理。下面通过示例演示 PL/pgSQL 提速的过程。

```
create or replace function slow_pi()
returns double precision
as $$
    declare
        a double precision := 1;
        s double precision := 1;
        r double precision := 0;
    begin
        for i in 1 .. 10000000 loop
            r := r + s/a;
            a := a + 2;
            s := -s;
        end loop;
        return 4 * r;
    end;
$$ language plpgsql;
```

上面的代码中采用了一种低效的方式模拟计算圆周率的过程，下面分别在 PostgreSQL 12 和 PostgreSQL 13 中进行测试。

在 PostgreSQL 12 中的执行时间大约是 13s。

```
postgres=# SELECT slow_pi();
     slow_pi
--------------------
 3.1415925535897915
(1 row)

Time: 13060.650 ms (00:13.061)
```

在 PostgreSQL 13 中的执行时间大约是 2s。

```
postgres=# select slow_pi();
     slow_pi
--------------------
```

```
 3.1415925535897915
(1 row)

Time: 2108.464 ms (00:02.108)
```

对比执行时间可以看出，循环代码在 PostgreSQL 13 中的执行速度得到了提高，更妙的是，循环代码无须进行任何修改，PL/pgSQL 即可比以前的执行速度更快。

4.1.7 Windows 连接优化

在 Windows 上，通常只能选择 TCP/IP 连接，TCP/IP 连接比 SOCKET 连接有更高的延迟。不过从 PostgreSQL 13 开始，Windows 开始支持本地 SOCKET 连接了。

在 Windows 上使用如下命令进行本地 SOCKET 连接。

```
psql -h c:\work\postgres\13\data -U postgres
```

通过执行\conninfo 命令可以看到，建立的是本地 SOCKET 连接。

```
postgres=# \conninfo
以用户 "postgres" 的身份，通过套接字"c:\work\postgres\13\data"在端口"5432"连接到数据库
  "postgres"
```

4.2 PostgreSQL 13 的可靠性提高

PostgreSQL 13 的可靠性提高主要体现在备份可靠性提高、流复制可动态配置。

4.2.1 备份可靠性提高

许多用户都用过 PostgreSQL 中的一些备份工具，应该了解备份在恢复之前很难验证有效性。PostgreSQL 13 引入了两个非常有用的特性来提高物理备份的可靠性，分别是备份清单 backup manifests 及备份验证工具 pg_verifybackup。

- backup manifests：通过 pg_basebackup 工具发起物理备份时列出的备份清单。
- pg_verifybackup：根据备份清单匹配验证备份内容的工具。

在 PostgreSQL 13 中使用 pg_basebackup 工具执行物理全备时会自动生成备份清单。备份完成之后，在备份目录下会存在一个 backup_manifest 文件，它是一个 JSON 数据类型的对象。

- PostgreSQL-Backup-Manifest-Version：备份清单的版本。
- Files：包含的文件列表，以及每个文件的相对路径，来自 PGDATA 和重要的元数据，如大小、最后修改时间和校验和。
- WAL-Ranges：时间线、备份开始的 LSN、备份结束的 LSN。
- Manifest-Checksum：校验和。

backup_manifest 文件的示例如下。

```
{ "PostgreSQL-Backup-Manifest-Version": 1,
```

```
"Files": [
{ "Path": "backup_label", "Size": 226, "Last-Modified": "2020-06-06 03:13:02 GMT",
  "Checksum-Algorithm": "CRC32C", "Checksum": "9c1588c0" },
{ "Path": "global/1262", "Size": 8192, "Last-Modified": "2020-06-02 09:47:56 GMT",
  "Checksum-Algorithm": "CRC32C", "Checksum": "a806d90f" },
…
],
"WAL-Ranges": [
{ "Timeline": 1, "Start-LSN": "0/6F000028", "End-LSN": "0/6F000138" }
],
"Manifest-Checksum":
  "6241b42fdc594dbf23c48390b9e204b6599d4cb6cef420efc374034e4d5a972b"}
```

pg_verifybackup 工具可以根据备份清单对物理全备文件验证备份的完整性。

下面是一个使用示例。

使用 pg_basebackup 工具进行物理全备。

```
$ pg_basebackup -D /backup/basebackup/test01 -Fp -Pv
pg_basebackup: initiating base backup, waiting for checkpoint to complete
pg_basebackup: checkpoint completed
pg_basebackup: write-ahead log start point: 0/71000028 on timeline 1
pg_basebackup: starting background WAL receiver
pg_basebackup: created temporary replication slot "pg_basebackup_46901"
353819/353819 kB (100%), 1/1 tablespace
pg_basebackup: write-ahead log end point: 0/71000138
pg_basebackup: waiting for background process to finish streaming ...
pg_basebackup: syncing data to disk ...
pg_basebackup: renaming backup_manifest.tmp to backup_manifest
pg_basebackup: base backup completed
```

检查生成的备份清单。

```
$ cd /backup/basebackup/test01/

$ ll backup_manifest
-rw------- 1 postgres postgres 262278 Aug 26 17:36 backup_manifest
```

使用 pg_verifybackup 工具进行验证。

```
$ pg_verifybackup /backup/basebackup/test01
backup successfully verified
```

如果删除 base 子目录下的某个数据文件，那么模拟数据将被损坏。

```
$ rm base/13580/16959
```

再次使用 pg_verifybackup 工具进行验证。

```
$ pg_verifybackup /backup/basebackup/test01
backup successfully verified
pg_verifybackup: error: "base/13580/16959" is present in the manifest but not on
  disk
```

从以上代码中可以看到明显的错误提示，备份清单发生了缺失。

使用 pg_verifybackup 工具验证备份完整性的步骤如下。

（1）检查备份目录。

- 忽略备份清单文件。
- 忽略 pg_wal 目录，通过单独的机制验证。

- 忽略 postgresql.auto.conf 文件、recovery.signal 文件和 standby.signal 文件，这些文件可能会更改。

（2）查找并解析备份清单。

默认情况下在备份目录中查找备份清单，使用-m, --manifest-path=PATH 选项可以指定备份清单的位置。在解析备份清单时，任何错误都被视为致命错误。

（3）扫描备份清单记录的文件。

将备份清单记录的文件和磁盘上的文件进行比较，若文件存在且大小匹配，则会在备份清单记录的相应文件处设置匹配标志，若没有设置匹配标志则会提示相应的错误。

（4）验证校验和。

在验证校验和时，会逐块读取备份清单文件，同时更新校验和，这需要付出较大的精力，在测试环境下可以使用-s,--skip-checksums 选项跳过这个步骤。

（5）验证并解析 WAL 记录。

备份清单中包含 WAL 记录的信息，pg_verifybackup 工具将使用该信息调用 pg_waldump 工具来解析 WAL 记录，在默认情况下，会在 pg_wal 目录下查找 WAL 记录，可以使用-w,--wal-directory=PATH 选项指定其他路径，如果没有备份 WAL 记录，那么可以使用-n,--no-parse-wal 选项跳过这个步骤。

（6）验证结束。

4.2.2 流复制可动态配置

PostgreSQL 13 之前的版本在更改流复制参数 primary_conninfo 和 primary_slot_name 时需要重启服务器，从 PostgreSQL 13 开始这两个参数的上下文生效环境由 postmaster 变为 sighup，允许动态配置后通过重新加载来生效。该特性避免了主库发生故障后重启备库的需要。

4.3 PostgreSQL 13 的运维管理优化

PostgreSQL 13 的运维管理优化主要体现在数据库删除更便捷、并行查询关联 PID、共享内存可观测、基于磁盘的缓存可监控、后台操作进度报告引入、语句日志采样降噪、PSQL 工具跟踪事务运行状态、pg_rewind 工具优化。

4.3.1 数据库删除更便捷

在 PostgreSQL 13 之前的版本中，如果有活动会话连接需要被删除的数据库，那么数据库的 DELETE 操作将不被允许进行，对多数场景来说，其实很难满足没有活动会话连接的情况，通常一直都会有新连接不断建立。通过修改数据库设置来阻塞新连接，同时删除已存在的连接，这样处理有点儿粗暴。

PostgreSQL 13 对数据库的 DELETE 操作提供了强制删除选项，在使用强制删除选项时，系统会先自动结束目标数据库的活动会话，再删除数据库。为了结束目标数据库的活动会话，删除数据库的用户需要有相应的操作权限，如 pg_terminate_backend 函数执行权限。但如果目标数据库中存在预备事务、活动的逻辑复制槽或订阅，那么 DELETE 操作会失败。

下面通过示例进行说明，直接删除目标数据库，提示目标数据库有活动会话正在使用，DELETE 操作会失败。

```
postgres=# DROP DATABASE testdb;
ERROR: database "testdb" is being accessed by other users
DETAIL: There is 1 other session using the database.
```

当使用 with(force) 选项时，系统会自动结束目标数据库中的活动会话，DELETE 操作更加友好。

```
postgres=# DROP DATABASE testdb WITH (force);
DROP DATABASE
```

此时目标数据库中的其他活动会话在下次执行代码时会提示如下信息。

```
FATAL:  terminating connection due to administrator command
server closed the connection unexpectedly
    This probably means the server terminated abnormally
    before or while processing the request.
The connection to the server was lost. Attempting reset: Failed.
```

删除数据库的另外一种方式是使用 dropdb 工具。

```
$ dropdb --force testdb
```

4.3.2 并行查询关联 PID

PostgreSQL 9.6 开始支持并行查询，并行查询会产生两种进程：worker 进程和 leader 进程。在活动会话视图的活动连接列表中很难看清进程之间的关联信息。PostgreSQL 13 在活动会话视图中新增了 leader_pid 字段，leader 进程的 leader_pid 字段的值为空，worker 进程的 leader_pid 字段的值为 leader 进程的 pid 字段的值。

使用多个并行进程对 test 表执行下面的代码。

```
explain analyze select sum(id) from test;
```

打开另一个窗口执行下面的代码，可以从 leader_pid 字段中清晰地看到进程之间的关联信息。

```
postgres=# select pid, leader_pid, query from pg_stat_activity where query like
   '%test%' and pid!=pg_backend_pid();
 pid  | leader_pid |                  query
------+------------+------------------------------------------
 4909 |            | explain analyze select sum(id) from test;
 5215 |       4909 | explain analyze select sum(id) from test;
 5216 |       4909 | explain analyze select sum(id) from test;
 5217 |       4909 | explain analyze select sum(id) from test;
 5218 |       4909 | explain analyze select sum(id) from test;
(5 rows)
```

也可以执行下面的代码进行更简化的输出。

```
postgres=# select query, leader_pid,
  array_agg(pid) as members
from pg_stat_activity
where leader_pid is not null
group by query, leader_pid;
                query                    | leader_pid |      members
-----------------------------------------+------------+---------------------
 explain analyze select sum(id) from test; |       4909 | {5215,5216,5217,5218}
(1 row)
```

4.3.3 共享内存可观测

PostgreSQL 13 新增了用于显示服务器主共享段分配信息的 pg_shmem_allocations 视图。其主要包括 postgres 进程本身分配的共享内存，以及通过参数 shared_preload_libraries 指定的外部扩展共享库使用的内存。

pg_shmem_allocations 视图中的字段说明如下。

- name：共享内存分配的名称，未使用的可用内存的 name 字段的值为 NULL，匿名分配的内存的 name 字段的值为 anonymous。
- off：偏移量，匿名分配的内存的 off 字段的值为 NULL。
- size：以字节为单位分配的大小。
- allocated_size：以字节为单位（包含填充）分配的大小。对于匿名分配的内存，由于没有填充相关的信息，因此 size 字段的值和 allocated_size 字段的值将始终相等。另外，对于未使用的可用内存的 size 字段的值和 allocated_size 字段的值也相等。

在默认情况下，只有具有超级用户权限才能查询 pg_shmem_allocations 视图，普通用户要查询该视图需要管理员赋予相关权限。

4.3.4 基于磁盘的缓存可监控

PostgreSQL 采用 SLRU（Simple Least-Recently-Used）缓存算法访问一些基于磁盘的信息。PostgreSQL 13 新增了 pg_stat_slru 视图，可以跟踪这些信息，每个基于磁盘的信息都会包含一行 SLRU 缓存算法的统计数据。

pg_stat_slru 视图中的字段说明如下。

- name：SLRU 缓存算法的名称。
- blks_zeroed：初始化期间被设置为零的块数。
- blks_hit：不需要读取磁盘块的次数。
- blks_read：读取磁盘块的次数。
- blks_written：写入磁盘块的次数。
- blks_exists：检测磁盘块存在的次数。
- flushes：刷脏的次数。
- truncates：截断的次数。
- stats_reset：统计数据最后一次重置的时间。

4.3.5 后台操作进度报告引入

PostgreSQL 13 新引入了两个后台操作进度报告，分别为 pg_stat_progress_basebackup 视图和 pg_stat_progress_analyze 视图。

目前，PostgreSQL 13 及 PostgreSQL 13 之前的版本支持的后台操作进度报告如表 4-1 所示。

表 4-1 PostgreSQL 13 及 PostgreSQL 13 之前的版本支持的后台操作进度报告

进度报告名称	描述	版本
pg_stat_progress_basebackup	实时跟踪使用 pg_basebackup 工具备份的各个阶段	PostgreSQL 13
pg_stat_progress_analyze	跟踪 ANALYZE 命令执行的进度	PostgreSQL 13
pg_stat_progress_create_index	在创建或重建索引时，显示当前创建索引的各个阶段的信息	PostgreSQL 12
pg_stat_progress_cluster	显示当前操作正在执行的各个阶段的信息	PostgreSQL 12
pg_stat_progress_vacuum	显示清理操作正在执行的各个阶段的信息	PostgreSQL 9.6

4.3.6 语句日志采样降噪

在一般情况下，用户很难把握慢语句执行时间的阈值，若某条比较繁忙的语句的执行时间超过了阈值，则数据库日志中将存在大量此条语句的日志，这个日志量可能会很惊人。在 PostgreSQL 13 中，可以使用参数 log_min_duration_sample 和 log_statement_sample_rate 来设置语句的基本采样值。

```
log_min_duration_sample = 100ms
log_statement_sample_rate = 0.2
```

上述语句的执行时间只有超过 100ms 才会开始记录日志，并且只会记录 20% 的日志。

当然，也可以与参数 log_min_duration_statement 一起使用。

```
log_min_duration_statement = 500ms
log_min_duration_sample = 100ms
log_statement_sample_rate = 0.2
```

由于参数 log_min_duration_statement 的优先级更高，上述语句的执行时间超过 500ms 会全部记录，执行时间在 100ms 与 500ms 之间则会记录 20% 的日志，因此合理的设置是参数 log_min_duration_sample 的值比参数 log_min_duration_statement 的值小。

4.3.7 PSQL 工具跟踪事务运行状态

PostgreSQL 的 PSQL 工具有 3 个客户端提示符变量。
- PROMPT1：等待新命令时发出的常规提示符。
- PROMPT2：输入了部分命令而没有全部完成，如在没有使用分号终止或引用没有被关闭时，等待更多输入时发出的提示符，PSQL 工具期待继续输入。

- PROMPT3：在运行 COPY FROM STDIN 并且需要在终端输入行值时发出的提示符。

变量 PROMPT2 的默认值与提示符变量 PROMPT1 一样，都为%/%R%#，PostgreSQL 13 对这两个提示符变量的值增加了%x，用于跟踪事务运行状态。

- 若不在事务中，则显示空字符。
- 若在事务中，则显示"*"。
- 若事务失败，则显示"!"。
- 若事务状态不确定，则显示"?"。

为了更好地理解，下面通过示例进行说明。

示例一：默认事务自动提交，不在事务中。

```
$ psql
psql (13.7)
Type "help" for help.

postgres=#
```

示例二：开启事务，显示"*"。

```
postgres=# begin;
BEGIN

postgres=*# select 1;
 ?column?
----------
        1
(1 row)
```

示例三：事务失败，显示"!"。

```
postgres=*# select 1 / 0;
ERROR: division by zero

postgres=!#

postgres=!# commit;
ROLLBACK
```

示例四：提示符变量 PROMPT2 的值增加了%w，可以生成与语句对齐的空格。

```
postgres=# \set PROMPT2 '%w'

postgres=# select 'a',
           'b',
           'c';
 ?column? | ?column? | ?column?
----------+----------+----------
 a        | b        | c
(1 row)
```

可以看到，输入的第 3 行和第 4 行语句的可读性更强，对于复杂语句更便于进行 COPY 操作。

4.3.8　pg_rewind 工具优化

在 PostgreSQL 主备环境下，如果在没有关闭主库的情况下将备库升为主库，那么主库和备库可能都在提供写入操作。在想将原主库以 standby 模式的角色重新加入集群，但是此时主库和备库之间的时间线已经偏离时，就需要使用 pg_rewind 工具。pg_rewind 工具相较于 pg_basebackup 工具的优势在于，pg_rewind 工具不需要读取数据库中未更改的数据块，如果数据库内容很大但只有少量的数据块发生改变，那么使用 pg_rewind 工具同步数据的速度会更快。

PostgreSQL 13 对 pg_rewind 工具新增了几个非常实用的附加功能。

1．-R / --write-recovery-conf 选项

使用这个选项，可以自动创建与流复制相关的恢复配置文件，并将指定的--source-server 选项中的连接字符串附加到 postgresql.auto.conf 文件的参数 primary_conninfo 中，这个选项可以用来将原主库快速恢复为备库。

2．-c / --restore-target-wal 选项

在使用 pg_rewind 工具进行崩溃恢复时，pg_wal 目录下的 WAL 文件可能会因某些原因不存在，而出现如下错误提示。

```
$ pg_rewind -D /var/lib/pgsql/data --source-server='host=node1 dbname=postgres
    user=postgres port=5432'
pg_rewind: servers diverged at WAL location 0/30005C8 on timeline 1
(... snip log output from postgres starting up in single user mode ...)
pg_rewind: error: could not open file
    "/var/lib/pgsql/data/pg_wal/000000010000000000000002": No such file or directory
pg_rewind: fatal: could not find previous WAL record at 0/2000100
```

当出现这种情况时，可以使用配置参数 restore_command 的命令来获取所需的 WAL 文件。

```
$ pg_rewind -D /var/lib/pgsql/data --source-server='host=node1 dbname=postgres
    user=postgres port=5432' --restore-target-wal
pg_rewind: servers diverged at WAL location 0/30005C8 on timeline 1
pg_rewind: rewinding from last common checkpoint at 0/2000060 on timeline 1
pg_rewind: Done!
```

3．自动恢复

pg_rewind 工具只能对干净关闭的数据库实例进行操作，否则它不能正确判断有哪些变更需要进行应用回放。在 PostgreSQL 13 中，当使用 pg_rewind 工具侦测到数据库实例未被干净关闭时，会自动以单用户模式进行崩溃恢复。

使用 pg_rewind 工具自动以单用户模式进行崩溃恢复的简化过程如下。

```
$ pg_rewind -D /var/lib/pgsql/data --source-server='host=node1 dbname=postgres
    user=postgres port=5432'
pg_rewind: executing "/usr/bin/pgsql/postgres" for target server to complete crash
    recovery
...
pg_rewind: servers diverged at WAL location 0/30005D0 on timeline 1
```

```
pg_rewind: rewinding from last common checkpoint at 0/2000060 on timeline 1
pg_rewind: Done!
```

4.4 PostgreSQL 13 的开发易用性提升

PostgreSQL 13 的开发易用性提升主要体现在分区表及逻辑复制改进、标识列可忽略用户输入、存储列可转换为普通列、分页排序可并列排名、易用的内置函数引入、FF1～FF6 时间格式引入、Row 表达式使用、视图列名纠正。

4.4.1 分区表及逻辑复制改进

在 PostgreSQL 13 中，对分区表进行了如下几个方面的改进。
- 允许在分区表上建立行级 BEFORE 触发器。
- 允许将整个分区表作为分区键。
- 可以直接显式锁定分区表（分区权限检测优化）。

逻辑复制在 PostgreSQL 13 之前的版本中只支持普通表，不支持分区表，如果需要对分区表进行逻辑复制，那么需要单独对所有分区进行逻辑复制。PostgreSQL 13 中的逻辑复制新增了对分区表的支持，包括以下 3 种情况。
- 将分区表复制到普通表中。
- 将分区表复制到异构分区表中。
- 将普通表复制到分区表中。

例如，创建分区表的 3 个分区，即 1 个历史分区，2 个按年分区。

```
CREATE TABLE tab_part1 (
  id serial,
  user_id int4,
  create_time timestamp(0) without time zone
) PARTITION BY RANGE(create_time);

CREATE TABLE tab_part1_his PARTITION OF tab_part1 FOR VALUES FROM (minvalue) TO
    ('2019-01-01');

CREATE TABLE tab_part1_2019 PARTITION OF tab_part1 FOR VALUES FROM ('2019-01-01')
    TO ('2020-01-01');

CREATE TABLE tab_part1_2020 PARTITION OF tab_part1 FOR VALUES FROM ('2020-01-01')
    TO ('2021-01-01');
```

在 PostgreSQL 13 之前的版本中如果创建发布时指定的表是分区表，那么会出现如下错误提示。

```
postgres=# CREATE PUBLICATION pub1 FOR TABLE tab_part1 ;
ERROR: "tab_part1" is a partitioned table
DETAIL: Adding partitioned tables to publications is not supported.
```

```
HINT:  You can add the table partitions individually.
```
PostgreSQL 13 可以直接发布分区表，并自动发布所有分区。
```
postgres=# CREATE PUBLICATION pub1 FOR TABLE tab_part1 ;
CREATE PUBLICATION

postgres=# select * from pg_publication_tables;
 pubname | schemaname |    tablename
---------+------------+-----------------
 pub1    | public     | tab_part1_his
 pub1    | public     | tab_part1_2019
 pub1    | public     | tab_part1_2020
(3 rows)
```
为了保证分区表数据复制正常，需要在目标端创建相同结构的分区表。如果目标端的分区策略不一样或目标端需要使用普通表，那么可以通过 PostgreSQL 13 新增的 publish_via_partition_root 选项进行设置。
```
postgres=# CREATE PUBLICATION pub2 FOR TABLE tab_part1
   with(publish_via_partition_root=true);
CREATE PUBLICATION

postgres=# select * from pg_publication_tables;
 pubname | schemaname | tablename
---------+------------+-----------
 pub2    | public     | tab_part1
(1 row)
```
设置 publish_via_partition_root 选项为 true 后，如果在创建发布中包含对分区表的更改，那么会使用分区父表的标识和结构进行发布，而不会使用默认各个分区的标识和结构进行发布。

在将普通表发布到目标端的分区表中时无须额外设置，与以前发布订阅的流程一样，只需要在订阅端创建好分区表结构即可。

4.4.2 标识列可忽略用户输入

标识列是 PostgreSQL 10 引入的功能，标识列支持使用 generated by default as identity 和 generated always as identity 两种形式控制用户是否可以显式赋值。

在创建表时，对于使用 generated by default as identity 创建的标识列字段，用户能够直接显式覆盖系统赋值。对于使用 generated always as identity 创建的标识列字段，用户不能直接显式覆盖系统赋值，但可以使用 overriding system value 覆盖系统赋值，否则插入或修改语句会报错。如果既不想让用户的赋值语句报错，又不想接收用户的显式赋值而保持系统赋值，那么可以使用 PostgreSQL 13 新增的 overriding user value。

下面的 test_identity2 表中的标识列使用了 generated always as identity。
```
create table test_identity2 (
   id int generated always as identity primary key,
   info text
);
```

如果想让用户输入的值 99 生效，那么可以使用 overriding system value 覆盖系统赋值。
```
postgres=# insert into test_identity2(id, info) overriding system value
 values (99, 'abc') returning *;
 id | info
----+------
 99 | abc
(1 row)
```
如果要丢弃用户赋值而使用系统赋值，那么可以使用 overriding user value。
```
postgres=# insert into test_identity2(id, info) overriding user value
 values (99, 'abc') returning *;
 id | info
----+------
  1 | abc
(1 row)
```

4.4.3　存储列可转换为普通列

Generated Columns 属性的值是依赖其他字段进行表达式计算得到的，使用 PostgreSQL 13 之前的版本无法删除字段的 Generated Columns 属性。在 PostgreSQL 13 中通过 ALTER TABLE 命令的 DROP EXPRESSION 选项可以删除字段的 Generated Columns 属性，示例如下。

```
CREATE TABLE t_test (
 a int,
 b int,
 c int GENERATED ALWAYS AS (a * b) STORED
);

INSERT INTO t_test (a, b) VALUES (10, 20);
```

创建测试表及插入测试数据之后，可以看到 c 字段的值通过表达式被自动生成了。
```
postgres=# SELECT * FROM t_test;
 a  | b  |  c
----+----+-----
 10 | 20 | 200
(1 row)
```

如果要显式定义 c 字段，那么只需要删除表达式即可。
```
postgres=# ALTER TABLE t_test ALTER COLUMN c DROP EXPRESSION ;
ALTER TABLE
```

通过查看下面的表结构可以发现，c 字段已经转换为普通字段了。
```
postgres=# \d t_test
           Table "public.t_test"
 Column |  Type   | Collation | Nullable | Default
--------+---------+-----------+----------+---------
 a      | integer |           |          |
 b      | integer |           |          |
 c      | integer |           |          |
```

此外，已有的 c 字段的值不会被删除，仍然存在。
```
postgres=# SELECT * FROM t_test;
```

```
 a  | b  |  c
----+----+-----
 10 | 20 | 200
(1 row)
```

4.4.4　分页排序可并列排名

在 PostgreSQL 13 中分页排序有如下两种形式。
- 形式一如下。
```
ORDER BY expression
LIMIT { count | ALL }
OFFSET start
```
- 形式二如下。
```
ORDER BY expression
OFFSET start { ROW | ROWS }
FETCH { FIRST | NEXT } [ count ] { ROW | ROWS } ONLY
```

下面通过示例演示分页排序的过程。

创建测试表，并插入数据。
```
CREATE TABLE test(name varchar,score int2);

INSERT INTO test(name, score) VALUES('张三','100'),('李四','90'),('王五1','85'),('王
    五2','85'),('王五3','85');
```

在 PostgreSQL 13 中，通常使用形式一进行分页排序，其中 LIMIT 用于控制返回的行数，OFFSET 用于控制起始行。
```
postgres=# \set PROMPT2 '%w'

postgres=# SELECT *
           FROM test
           ORDER BY score DESC
           LIMIT 2 OFFSET 0;
 name | score
------+-------
 张三 |   100
 李四 |    90
(2 rows)
```

也可以使用形式二进行分页排序，其中 OFFSET 用于控制起始行，FETCH 用于控制返回的行数。
```
postgres=# select *
           FROM test
           ORDER BY score DESC
           OFFSET 0
           FETCH FIRST 2 ROWS ONLY;
 name | score
------+-------
 张三 |   100
 李四 |    90
```

```
(2 rows)
```

在 PostgreSQL 13 中，对形式二增加了 WITH TIES 选项，具体如下。

```
FETCH { FIRST | NEXT } [ count ] { ROW | ROWS } { ONLY | WITH TIES}
```

使用 WITH TIES 选项可以解决排名相同的问题，示例如下。

```
postgres=# SELECT *
        FROM test
        ORDER BY score DESC
        OFFSET 0
        FETCH FIRST 3 ROWS WITH TIES;
 name | score
------+-------
 张三  |  100
 李四  |   90
 王五1 |   85
 王五2 |   85
 王五3 |   85
(5 rows)
```

其中，参数 FETCH FIRST 的值为 3，按照以前的形式应该返回 3 条数据，但是这里返回了 5 条数据。WITH TIES 选项用于设置返回排名和最后一行相同的数据行，分数为 85 分的 3 条数据全部被返回。

4.4.5 易用的内置函数引入

PostgreSQL 13 之前的版本不提供 UUID 数据类型的内置函数，如果需要使用 UUID 数据类型的内置函数，那么需要加载外部 uuid_ossp 模块或 pgcrypto 模块。从 PostgreSQL 13 开始，可以直接使用 gen_random_uuid 函数，示例如下。

```
postgres=# SELECT gen_random_uuid();
           gen_random_uuid
--------------------------------------
 960d6103-090e-472e-901e-daac7b73a3a3
(1 row)
```

min_scale 函数用于精确表示返回的最小数字，trim_scale 函数用于移除尾部的零。

```
postgres=# SELECT n, min_scale(n), trim_scale(n) FROM (VALUES(98.5600)) as t(n);
   n     | min_scale | trim_scale
---------+-----------+------------
 98.5600 |         2 |      98.56
(1 row)
```

gcd 函数用于计算最大公约数，lcm 函数用于计算最小公倍数。

```
postgres=# SELECT gcd(54,24), lcm(54,24);
 gcd | lcm
-----+-----
   6 | 216
(1 row)
```

4.4.6　FF1～FF6 时间格式引入

SQL 标准 2016 定义了 FF1～FF6 时间格式，PostgreSQL 13 引入了对 FF1～FF6 时间格式的支持，FF1～FF6 时间格式表示秒后的第 1 位到第 6 位，说明如下。
- FF1：秒的 1/10。
- FF2：秒的 1/100。
- FF3：秒的 1/1000，毫秒。
- FF4：毫秒的 1/10。
- FF5：毫秒的 1/100。
- FF6：毫秒的 1/1000，微秒。

在 PostgreSQL 13 中执行以下语句，可以通过 FF1～FF6 时间格式分别返回秒后的第 1 位到第 6 位。

```
postgres=# \x
Expanded display is on.

postgres=# SELECT now(),
       to_char(now(),'HH24:MI:SS.FF1'),
       to_char(now(),'HH24:MI:SS.FF2'),
       to_char(now(),'HH24:MI:SS.FF3'),
       to_char(now(),'HH24:MI:SS.FF4'),
       to_char(now(),'HH24:MI:SS.FF5'),
       to_char(now(),'HH24:MI:SS.FF6');
-[ RECORD 1 ]--------------------------
now     | 2020-09-04 14:48:01.022295+08
to_char | 14:48:01.0
to_char | 14:48:01.02
to_char | 14:48:01.022
to_char | 14:48:01.0222
to_char | 14:48:01.02229
to_char | 14:48:01.022295
```

此外，FF1～FF6 时间格式还可用于 to_timestamp 函数及 jsonpath 查询的 .datetime 函数。

4.4.7　Row 表达式使用

Row 表达式是一个行构造器，PostgreSQL 13 允许使用后缀列名提取列成员，示例如下。第一列通过列名 f1 提取列成员，第二列通过列名 f2 提取列成员。

```
postgres=# select (row(100,'abc'::varchar,3.14)).f1;
 f1
------
 100
(1 row)
```

4.4.8 视图列名纠正

在 PostgreSQL 13 之前的版本中，要修改视图列名必须重新创建视图，示例如下。

```
CREATE VIEW uptime AS SELECT current_timestamp, current_timestamp -
  pg_postmaster_start_time();
```

从以上代码中可以看出，在视图的第二列，忘记了指定别名，可以使用下面的代码纠正。

```
ALTER VIEW uptime RENAME COLUMN "?column?" TO uptime;
```

4.5 PostgreSQL 13 的系统层变化

PostgreSQL 13 的系统层变化主要体现在系统元数据引入、配置参数引入、对象标识符类型引入、内部术语引入、备库升主库流程优化、INSERT 操作自动清理调优，WAL 用量跟踪、坏块绕过继续恢复、外部表安全性提高，以及附加模块变化。

4.5.1 系统元数据引入

PostgreSQL 13 新引入了 4 个系统元数据，如表 4-2 所示。

表 4-2 PostgreSQL 13 新引入的系统元数据

元数据名称	描述
pg_shmem_allocations	显示主共享段内存分解的使用信息
pg_stat_progress_analyze	跟踪 ANALYZE 操作执行的状态进度
pg_stat_progress_basebackup	实时跟踪使用 pg_basebackup 工具备份的各个阶段
pg_stat_slru	显示 SLRU 缓存算法的统计信息

4.5.2 配置参数引入

PostgreSQL 13 新引入了 16 个配置参数，如表 4-3 所示。

表 4-3 PostgreSQL 13 新引入的配置参数

参数名称	描述
hash_mem_multiplier	基于 HASH 操作使用的最大内存量
logical_decoding_work_mem	逻辑解码使用的最大内存量
maintenance_io_concurrency	与参数 effective_io_concurrency 相似，但用于支持多个客户端会话完成的维护工作
wal_skip_threshold	当参数 wal_level 的值为 minimal 并且在 create 或 rewriting 一个永久关系之后进行事务提交时，本参数用于设置如何持久化新数据
wal_keep_size	用于指定流复制环境下为备库保留 WAL 文件的最小尺寸
max_slot_wal_keep_size	用于限制复制槽保留 WAL 文件的最大尺寸

续表

参数名称	描述
wal_receiver_create_temp_slot	在流复制环境下，如果在备库中未设置参数 primary_slot_name，那么在备库中设置本参数的值为 true 之后，可以在主库创建临时复制槽
enable_incremental_sort	用于设置启用或禁用增量排序
log_min_duration_sample	日志抽样开关，只有达到某个值才开始抽样记录
log_statement_sample_rate	用于设置日志抽样记录的比例
log_parameter_max_length	用于设置在语句执行正确时绑定参数值允许的最大字节数
log_parameter_max_length_on_error	用于设置在语句执行错误时绑定参数值允许的最大字节数
autovacuum_vacuum_insert_threshold	表只有 INSERT 操作触发 VACUUM 操作的元组数
autovacuum_vacuum_insert_scale_factor	表只有 INSERT 操作触发 VACUUM 操作的元组比例
backtrace_functions	调试选项，用于设置在运行错误匹配设置的内置函数时，服务端文件对应跟踪的错误信息
ignore_invalid_pages	恢复选项，用于设置在数据库恢复过程中忽略遇到的坏块并继续恢复

4.5.3 对象标识符类型引入

对象标识符（OID）用于内部标识。使用它的一些类型可以方便用户进行关联查询。PostgreSQL 13 对 Collation 类型增加了 regcollation 的别名支持。在 PostgreSQL 13 中可以使用 to_regcollation 函数。

```
postgres=# select '"POSIX"'::regcollation;
 regcollation
--------------
 "POSIX"
(1 row)

postgres=# select to_regcollation('"POSIX"');
 to_regcollation
-----------------
 "POSIX"
(1 row)
```

PostgreSQL 13 新引入的对象标识符类型如表 4-4 所示。

表 4-4　PostgreSQL 13 新引入的对象标识符类型

对象标识符名称	类型
regclass	pg_class
regcollation	pg_collation
regconfig	pg_ts_config
regdictionary	pg_ts_dict
regnamespace	pg_namespace
regoper	pg_operator
regoperator	pg_operator

续表

对象标识符名称	类型
regproc	pg_proc
regprocedure	pg_proc
regrole	pg_authid
regtype	pg_type

4.5.4 内部术语引入

PostgreSQL 13 官方文档的附录部分引入了一个附录 M 的术语表，涵盖了 PostgreSQL 13 中的一些基础信息。下面列举几个常用的内部术语。

- Tuple：元组，是大部分人理解的表中的行。它由一些列组成，每列由列名、列值和数据类型组成。人们经常听到的 Tuple Set，即行集，也就是常说的表，或类似于表的对象。
- Record 与 Row：记录与行，Record 与 Row 是 Tuple 的另一种说法，与 Tuple 不同的是，Record 与 Row 一般用于 SQL 语句中。Record 一般用于数据库服务端函数返回复合数据的类型。Row 一般用于在 SQL 语句中构建复合数据类型。
- Relation：关系或对象，即行的集合。通常出现的对象为表，每个表均是对象，反之不然。View（视图）、Index（索引）、Sequence（序列）也是对象的一种，对象为查询的结果，但查询的中间步骤也可以看作一个对象。
- Schema：模式，类似于数据库中的一个文件夹。实际上，PostgreSQL 13 内部使用的命名空间（namespace）指 Schema。
- Page：页，用于表示数据库内部的实现细节，对最终用户来说是透明的。DBA 需要对页有一定的理解，在磁盘层面，数据库表以文件形式存在。在一般情况下一个文件对应一个表（实际上当表的容量足够大时一个表对应多个文件）。如果表中的数据行都紧凑存储在一个文件中，那么当修改数据时，为了扩大空间，就需要大量移动数据。在数据库内部将表文件划分为间隔规则的页，并仅用元组填充每页的部分。这样当数据库需要更新一个元组时，只需要重写相关的页，而文件的其余部分可以保持不变。

4.5.5 备库升主库流程优化

在 PostgreSQL 12 及 PostgreSQL 12 之前的版本中，如果备库进行应用回放 WAL 被暂停，即 WAL 没有完成应用回放，那么此时备库执行 promote 操作升主库会失败。这是因为当暂停 WAL 应用回放时，PostgreSQL 12 及 PostgreSQL 12 之前的版本并不能响应备库升主库的请求，除非应用回放 WAL 暂停恢复并且完成所有待回放的 WAL。在这种情况下，如果尝试执行 promote 操作，那么 PostgreSQL 12 及 PostgreSQL 12 之前的版本将处于一个不可预期的时间点并且可能会执行超时。这个限制在 PostgreSQL 13 中得到了优化，promote 操作可以被优先执行。

4.5.6 INSERT 操作自动清理调优

很多人可能了解使用 AUTOVACUUM 进程处理死元组的必要性，出现死元组是 PostgreSQL 的多个版本并发控制产生的负面影响。从 PostgreSQL 13 开始，系统会对表收集自上次进行 VACUUM 操作之后已经插入的行数，用户可以通过 pg_stat_all_tables 视图，或 pg_stat_user_tables 视图和 pg_stat_sys_tables 视图的 n_ins_since_vacuum 字段观察该值。当该值超过下面两个新增参数设置的值时，会触发 AUTOVACUUM 进程（参数 reltuples 的值表示表的预估行数，存储于 pg_class 表中）。

```
autovacuum_vacuum_insert_threshold +
autovacuum_vacuum_insert_scale_factor * reltuples
```

PostgreSQL 13 对 insert-only 表增加了 VACUUM 操作，主要是为了处理以下几个问题。
- 事务环绕问题。
- 仅索引扫描问题。
- 提示位问题。

1．事务环绕问题

PostgreSQL 13 将事务 ID 存储在表的系统列 xmin 和 xmax 中，以此来确定哪个行版本对查询语句可见。事务 ID 是无符号的 4 字节整数，当超过大约 40 亿个事务之后，计数器所计的数将达到上限，需要"环绕"并从数字 3 重新开始计数。由于"环绕"循环使用，因此到大约 20 亿个事务之后，数据可能会丢失，在此之前表的现存数据行需要被冻结标识为无条件可见，这也是 AUTOVACUUM 进程的工作之一。

在 PostgreSQL 13 中可以使用上面新增的两个参数来避免受到此问题的影响，而对于 PostgreSQL 13 之前的版本，可以通过更早触发冻结来达到类似的效果。例如，可以设置每到 10 万个事务对表进行一次冻结，通过表的存储参数进行如下设置。

```
ALTER TABLE mytable SET (
autovacuum_freeze_max_age = 100000
);
```

2．仅索引扫描问题

虽然表中有系统列 xmin 和 xmax，可以确定哪个行版本对查询语句可见，但是索引不包含此信息。为了解决索引的可见性问题，PostgreSQL 13 引入了可见性映射的数据结构，该数据结构对表的每个数据块存储两个比特位，其中一个比特位用于标识数据块中的所有行是否对所有事务可见。如果索引扫描条目从可见性映射文件中发现包含引用的行，那么它可以跳过检查该条目的可见性。

为了让仅索引能通过可见性映射文件跳过回表操作，需要 VACUUM 操作运行得足够频繁。通过减小参数 autovacuum_vacuum_scale_factor 的值只能让表的 UPDATE 操作和 DELETE 操作及时触发 AUTOVACUUM 进程更新可见性映射文件，insert-only 表需要通过减小参数 autovacuum_vacuum_insert_scale_factor 的值进行触发。

```
ALTER TABLE mytable SET (
autovacuum_vacuum_insert_scale_factor = 0.005
);
```

PostgreSQL 13 之前的版本可以通过定时任务手动发起 VACUUM 操作或通过减小参数 autovacuum_freeze_max_age 的值来使 AUTOVACUUM 进程运行得足够频繁。

3. 提示位问题

在 PostgreSQL 13 的新创建行的首次查询中，必须通过查阅事务提交日志文件来确定事务是否已提交。在首次查询之后，会在该行上设置一个提示位，记录事务状态，这样之后的查询可以减少检查事务提交日志文件的工作，避免频繁访问事务提交日志文件进而影响性能。

如果表中插入大量的行，那么第一次查询请求可能会因更新提示位信息使得性能下降。因此，当表执行了大量的 INSERT 操作或 COPY 操作之后，比较好的习惯是再次对表执行 VACUUM 操作。

然而，在很多情况下并不总遵循这个最佳实践，而需要手动发起额外操作又确实比较烦人。有了这个新特性，PostgreSQL 13 会在插入大量数据之后自动清理表。

4.5.7 WAL 用量跟踪

WAL 膨胀是运维过程中比较常见的问题，PostgreSQL 13 允许对 EXPLAIN 命令、auto_explain 模块、AUTOVACUUM 进程和 pg_stat_statements 模块跟踪 WAL 用量。

1. 对 EXPLAIN 命令跟踪 WAL 用量

EXPLAIN 命令可以通过 WAL 选项（同时依赖 ANALYZE 选项）跟踪 WAL 用量，WAL 的信息包括以下 3 个部分。

- records：WAL 记录数。
- fpi：产生全页写的数量。
- bytes：WAL 字节量。

下面通过示例演示 WAL 用量跟踪的过程。

创建测试表。

```
postgres=# CREATE TABLE tab_wal(id int,info text,create_time timestamp(0) without
    time zone);
CREATE TABLE
```

插入 100 万条数据。

```
postgres=# EXPLAIN (ANALYZE ON,WAL ON) INSERT INTO tab_wal(id,info,create_time)
    SELECT n,'test_'|| n,clock_timestamp() FROM generate_series(1,1000000) n;
                        QUERY PLAN
-----------------------------------------------------------------------
 Insert on tab_wal  (cost=0.00..35000.00 rows=1000000 width=44)
(actual time=4744.766..4744.768 rows=0 loops=1)
   WAL: records=1000000 bytes=79000008
   ->  Subquery Scan on "*SELECT*"  (cost=0.00..35000.00 rows=1000000 width=44)
    (actual time=282.350..1710.650 rows=1000000 loops=1)
         ->  Function Scan on generate_series n  (cost=0.00..20000.00 rows=1000000
    width=44) (actual time=282.347..1174.406 rows=1000000 loops=1)
 Planning Time: 0.095 ms
```

```
Execution Time: 4747.556 ms
(6 rows)
```

在上述代码中，records=1000000，表示和插入的记录数对应；bytes=79000008，表示 75MB 左右的 WAL，与 pg_wal 目录产生 5 个 16MB 的 WAL 对应。

修改表产生的 WAL 的信息如下。

```
postgres=# EXPLAIN (ANALYZE ON,WAL ON) UPDATE tab_wal SET info = info||'test';
                          QUERY PLAN
-----------------------------------------------------------------------
 Update on tab_wal  (cost=0.00..18870.00 rows=1000000 width=50)
 (actual time=8711.690..8711.693 rows=0 loops=1)
   WAL: records=2000000 bytes=159999208
   ->  Seq Scan on tab_wal  (cost=0.00..18870.00 rows=1000000 width=50) (actual
   time=0.009..548.140 rows=1000000 loops=1)
 Planning Time: 0.237 ms
 Execution Time: 8711.791 ms
(5 rows)
```

在上述代码中，records=2000000，由于使用 MVCC 机制，因此增加了一倍；bytes=159999208，表示 152MB 左右的 WAL，同样也是增加了大约一倍。

删除表产生的 WAL 的信息如下。

```
postgres=# EXPLAIN (ANALYZE ON,WAL ON) DELETE FROM tab_wal ;
                         QUERY PLAN
-----------------------------------------------------------------------
 Delete on tab_wal  (cost=0.00..23723.00 rows=1000000 width=6)
 (actual time= 2494.523..2494.524 rows=0 loops=1)
   WAL: records=1000000 bytes=54000000
   ->  Seq Scan on tab_wal  (cost=0.00..23723.00 rows=1000000 width=6)
 (actual time=4.283..252.838 rows=1000000 loops=1)
 Planning Time: 0.100 ms
 Execution Time: 2494.599 ms
(5 rows)
```

在上述代码中，records=1000000，表示和插入的记录数对应；bytes=54000000，表示 51MB 左右的 WAL。

2. 对 auto_explain 模块跟踪 WAL 用量

通过设置参数 auto_explain.log_wal 可以跟踪 WAL 用量，从数据库日志文件中可以观察到如下信息。

```
2020-07-05 15:04:07.767 CST [51405] LOG:  statement: update vac_test  set name =
    '111' where id = 10 ;
2022-07-05 15:04:07.767 CST [51405] LOG:  duration: 0.047 ms  plan:
        Query Text: update vac_test  set name = '111' where id = 10 ;
        Update on vac_test  (cost=0.00..1.25 rows=1 width=88) (actual
    time=0.046..0.047 rows=0 loops=1)
          WAL: records=1 fpi=1 bytes=1179
          ->  Seq Scan on vac_test  (cost=0.00..1.25 rows=1 width=88) (actual
    time=0.005..0.006 rows=1 loops=1)
                Filter: (id = 10)
                Rows Removed by Filter: 19
```

3. 对 AUTOVACUUM 进程跟踪 WAL 用量

通过设置参数 log_autovacuum_min_duration，同样可以从数据库日志文件中观察到 WAL 用量。

```
2020-06-02 12:02:48.230 CST [56372] LOG:  automatic vacuum of table
  "maleah_db.public.vac_test": index scans: 0
    pages: 0 removed, 1 remain, 0 skipped due to pins, 0 skipped frozen
    tuples: 20 removed, 20 remain, 0 are dead but not yet removable, oldest
xmin: 832
    buffer usage: 31 hits, 0 misses, 3 dirtied
    avg read rate: 0.000 MB/s, avg write rate: 47.541 MB/s
    system usage: CPU: user: 0.00 s, system: 0.00 s, elapsed: 0.00 s
    WAL usage: 6 records, 4 full page images, 33438 bytes
```

4. 对 pg_stat_statements 模块跟踪 WAL 用量

pg_stat_statements 模块的 pg_stat_statements 视图新增了 3 个字段，即 wal_records、wal_fpi、wal_bytes，用于跟踪 WAL 用量。

4.5.8 坏块绕过继续恢复

在 PostgreSQL 中有一个参数 zero_damaged_pages，用于控制 PostgreSQL 运行过程中遇到坏块时是否绕过这个坏块。当参数 zero_damaged_pages 的值为 on 时，系统在检测到受损页开头时不会报错，允许系统忽略损坏的数据页，将其视为全零的空页。当出现软件或硬件故障导致数据损坏时，该参数可用于恢复数据。

PostgreSQL 13 新增了参数 ignore_invalid_pages 用于控制数据库恢复过程中遇到坏块时是否绕过这个坏块。参数 ignore_invalid_pages 的默认值为 off，当在数据库恢复过程中发现 WAL 记录引用了无效页时，系统将出现严重错误，中止恢复。如果将参数 ignore_invalid_pages 的值设置为 on，那么在恢复过程中发现 WAL 记录引用了无效页时，数据库将忽略这个严重错误，只发出警告，并继续进行恢复，这种行为可能会导致崩溃、数据丢失、隐藏损坏或其他严重问题。当遇到这种情况时，应尽量尝试恢复已损坏的数据文件。

参数 zero_damaged_pages 与 ignore_invalid_pages 的功能的相同之处在于都用于控制数据库遇到坏块时是否恢复严重受损的数据，不同之处在于参数 zero_damaged_pages 用于数据库运行过程中遇到数据坏块的场景，参数 ignore_invalid_pages 用于数据库处于 recovery 模式或 standby 模式中遇到坏块的场景。

4.5.9 外部表安全性提高

对于 PostgreSQL 13 之前的版本，只有超级用户可以使用 postgres_fdw 模块与其建立免密连接，普通用户不能使用 trust 方法建立免密连接。PostgreSQL 13 引入了参数 password_required，允许普通用户建立免密连接。

下面分别使用超级用户权限及普通用户权限进行外部表查询。

```
create table fdw_test (id int);
insert into fdw_test values(1),(2);
create extension postgres_fdw;
create server myserver_test foreign data wrapper postgres_fdw options (dbname
    'postgres');
create foreign table fdw_svr(id int) server myserver_test options(table_name
    'fdw_test');
create user mapping for public server myserver_test;
```

当使用超级用户权限进行外部表查询时，可以正常返回结果。

```
postgres=# select * from fdw_svr;
 id
----
  1
  2
(2 rows)
```

当使用普通用户权限进行外部表查询时，会提示非超级用户，需要提供密码。

```
postgres=# create user nonsup;
CREATE ROLE

postgres=# grant all on fdw_svr to nonsup;
GRANT

postgres=# \c - nonsup
You are now connected to database "postgres" as user "nonsup".

postgres=> select * from fdw_svr;
ERROR: password is required
DETAIL: Non-superusers must provide a password in the user mapping.
```

在 PostgreSQL 13 中创建用户免密连接时可以使用参数 password_required。

```
postgres=# create user mapping for nonsup server myserver_test options
    (password_required 'false');
CREATE USER MAPPING

postgres=# \c - nonsup
You are now connected to database "postgres" as user "nonsup".

postgres=> select * from fdw_svr;
 id
----
  1
  2
(2 rows)
```

此时，普通用户也可以建立免密连接正常进行外部表查询。

4.5.10 附加模块变化

PostgreSQL 13 中附加模块的变化具体如下。

- 新增 trusted（可信标识），创建可信标识的模块不需要超级用户权限。
- auto_explain 模块新增参数 log_wal，用于跟踪 WAL 用量。
- pg_stat_statements 模块的 pg_stat_statements 视图新增 3 个字段，用于跟踪 WAL 用量。
    - wal_records：WAL 记录数。
    - wal_fpi：产生全页写的数量。
    - wal_bytes：WAL 字节量。

新增参数 track_planning，用于控制跟踪执行计划耗时信息。需要注意，打开这个参数会降低 SQL 语句的性能，这个参数默认是被关闭的，同时 pg_stat_statements 视图相应增加与执行计划耗时相关的参数。
- plans：被解析生成执行计划的次数。
- total_plan_time：生成执行计划的时间总和，单位为毫秒。
- min_plan_time：生成执行计划的最短时间，单位为毫秒。
- max_plan_time：生成执行计划的最长时间，单位为毫秒。
- mean_plan_time：生成执行计划的平均时间，单位为毫秒。
- stddev_plan_time：执行计划耗时的总体标准偏差，单位为毫秒。
- postgres_fdw 模块新增参数 password_required，允许普通用户建立免密连接。

4.6 本章小结

PostgreSQL 13 在性能提升方面主要包括索引去重，通过索引去重不仅可以减小存储空间，而且可以提高查询效率；利用多核特性可以并发重建库级索引；通过 HASH 聚合溢出到磁盘，可以降低内存溢出风险；通过并行清理索引，可以加快单表清理速度等。

PostgreSQL 13 在可靠性提高方面主要包括引入了备份清单及备份工具，备份工具可以根据产生的备份清单进行备份有效性验证等。

PostgreSQL 13 在运维管理优化方面，主要包括数据库可以自动结束已有的活动会话连接进行强制删除；新增共享内存及基于磁盘缓存的可观测视图；通过使用查询语句设置采样比例可以降低日志文件的输出量等。

PostgreSQL 13 在开发易用性提升方面主要包括分区表可以逻辑复制到普通表或异构分区表中；存储列及标识列引入了更丰富的功能；分页排序可并列排名；引入了更多易用的内置函数等。PostgreSQL 13 在系统层变化方面主要包括备库升主库流程优化；WAL 用量跟踪；坏块绕过继续恢复等。

第 5 章

PostgreSQL 14 新特性

5.1 PostgreSQL 14 的主要性能提升

与 PostgreSQL 14 之前的版本相比，PostgreSQL 14 的性能有了一定的提升，主要体现在高并发连接优化、紧急清理模式引入、列级压缩可配置、逻辑复制改进、嵌套循环改进、并行特性增强。

5.1.1 高并发连接优化

PostgreSQL 的连接模型依赖于进程，建立大量连接对性能影响较大。这个对性能的影响并不是由于连接过多导致内存使用增加，而是由于连接过多使得频繁计算事务快照的费用非常昂贵。在 PostgreSQL 14 中，优化了计算事务快照的 GetSnapshotData 函数，活动连接和空闲连接的扩展性得到了明显的改善。

假设使用两个 96 核 AWS 实例（c5.24xlarge），一个运行 PostgreSQL 13，一个运行 PostgreSQL 14，在 Ubuntu 20.04 系统下使用 pgbench 工具进行测试。

```
$ pgbench -i -s 200
```

```
$ pgbench -S -c 5000 -j 96 -M prepared -T30
```

在连接 5000 个活动时，PostgreSQL 14 的 TPS（Transactions Per Second）为 49.5 万，PostgreSQL 13 的 TPS 为 41.8 万。

提交高并发连接优化特性的原作者 Andres Freund 对连接单个活动的 TPS 进行了基准

测试，在保持 1 万个空闲连接的情况下，PostgreSQL 14 的 TPS 为 3.5 万，PostgreSQL 13 的 TPS 为 1.5 万。

5.1.2 紧急清理模式引入

在 PostgreSQL 中，VACUUM 操作大致需要经历下面几个主要阶段。
- 扫描堆，找到所有死元组并冻结。
- 进行索引清理，扫描所有索引并从每个索引中删除元组。
- 清理堆，清除索引，截断堆，将数据块末尾的空闲空间返还给操作系统。
- 更新空闲空间映射（VM）文件和 FSM 文件，以及更新统计信息等。

VACUUM 操作的这几个阶段相对比较耗时，在 PostgreSQL 14 中对 VACUUM 操作增加了一个紧急模式。当表的年龄超过参数 vacuum_failsafe_age 的值时会触发紧急模式，在紧急模式下，表的 VACUUM 操作会被强制加速，不会受 VACUUM 操作代价及延迟参数的限制，并且会跳过一些不必要的维护工作，如索引的清理工作等。此时，VACUUM 操作将以最快的速度冻结来更好地预防事务 ID 回卷。在子事务环境下，也有类似机制，通过参数 vacuum_multixact_failsafe_age 控制。

5.1.3 列级压缩可配置

在 PostgreSQL 中，页是数据存储的基本单位，页的大小默认为 8KB，数据行不允许跨页存储。然而，有些数据类型的长度是可变的，数据行有可能超过一页。为了克服这个限制，大的字段值被压缩或分解成多个物理行，这种技术被称为 The Oversized-Attribute Storage Technique，简称 TOAST。

在 PostgreSQL 14 之前的版本中，TOAST 只支持一种内置的压缩算法，即 pglz 算法，不能配置。PostgreSQL 14 新增了参数 default_toast_compression，默认值为 pglz，可配置的值为 pglz 和 lz4。

也可以对列进行显式配置，示例如下。

```
create table tab_compression(
a text compression pglz,
b text compression lz4
);
```

注意，数据库内核只支持 pglz 算法。要使用 lz4 算法，需要在编译 PostgreSQL 时打开 --with-lz4 选项，否则会出现下面的错误提示。

```
ERROR: compression method lz4 not supported
DETAIL: This functionality requires the server to be built with lz4 support.
HINT: You need to rebuild PostgreSQL using --with-lz4.
```

lz4 算法是一种以速度著称的无损压缩算法，可以有效地提高压缩和解压缩的速度，它的压缩性能比 pglz 算法更好，且使用更少的 CPU。

观察下面的测试可以发现，lz4 算法与 pglz 算法的性能相比有了很大的提升。

```
CREATE TABLE tab_compression_2 (
id int,
```

```
data text compression pglz
);
```

使用 pglz 算法对数据列进行压缩，插入 100 万条数据大约用时 67s。

```
postgres=# insert into tab_compression_2 select i, repeat('x',10000) from
    generate_series(1,1000000) i;
INSERT 0 1000000
Time: 66568.681 ms (01:06.569)
```

pglz 算法的空间占用 160MB。

```
postgres=# select pg_size_pretty(pg_total_relation_size('tab_compression_2'));
 pg_size_pretty
----------------
 160 MB
(1 row)
```

在清空数据后，通过 ALTER TABLE 命令修改 data 列的压缩算法为 lz4 算法。

```
truncate table tab_compression_2 ;
alter table tab_compression_2 alter column data set compression lz4;
```

在使用新的 lz4 算法压缩后，插入相同的数据大约用时 8s，效率提高近 10 倍。

```
postgres=# insert into tab_compression_2 select i, repeat('x',10000) from
    generate_series(1,1000000) i;
INSERT 0 1000000
Time: 7677.512 ms (00:07.678)
```

同时，lz4 算法的空间占用 89MB，比 pglz 算法节省了将近一倍空间。

```
postgres=# select pg_size_pretty(pg_total_relation_size('tab_compression_2'));
 pg_size_pretty
----------------
 89 MB
(1 row)
```

可以看出，与 pglz 算法相比，lz4 算法在压缩性能及空间占用方面有很大的优势。

数据列的压缩算法在运行过程中可能会被修改，示例如下。

```
truncate table tab_compression_2 ;
alter table tab_compression_2 alter column data set compression pglz;
insert into tab_compression_2 values(1, repeat('1234567890', 1000));
alter table tab_compression_2 alter column data set compression lz4;
insert into tab_compression_2 values(2, repeat('1234567890', 1000));
```

可以使用新增的 pg_column_compression 函数来查看数据列的实际压缩算法。

```
postgres=# select id, pg_column_compression(data) from tab_compression_2;
 id | pg_column_compression
----+-----------------------
  1 | pglz
  2 | lz4
(2 rows)
```

在使用 CREATE TABLE AS SELECT 命令或 SELECT INSERT 命令复制数据时也可能会存在多种压缩算法，示例如下。

```
postgres=# create table tab_compression_3 as select * from tab_compression_2;
SELECT 2

postgres=# select id, pg_column_compression(data) from tab_compression_3;
```

```
 id | pg_column_compression
----+-----------------------
  1 | pglz
  2 | lz4
(2 rows)
```

由此可知，建议使用参数 default_toast_compression 来对所有数据列配置统一的压缩算法，避免使用混合的压缩算法。与此相关的是，pg_dump 工具和 pg_dumpall 工具也添加了一个--no-toast-compression 选项，这样可以避免出现因导出列级压缩选项而覆盖目标数据库实例级别的全局配置问题。

5.1.4 逻辑复制改进

PostgreSQL 14 对逻辑复制进行了如下改进。

- 支持流式处理正在进行的大事务。

在 PostgreSQL 14 之前的版本中，逻辑复制仅在事务提交后进行解码和复制，这是为了避免最终可能中止的事务。PostgreSQL 14 新增了流式处理功能，无须等待事务完成，即可将解码后更新的信息传输到订阅端，这有助于降低大事务的回放延迟，带来较大的性能提升。

该功能是通过设置 streaming 选项来实现的，示例如下。

```
alter subscription mysub set (streaming = on);
```

- 提升 TRUNCATE 操作的性能。

在 PostgreSQL 14 之前的版本中，对有 1000 个分区的表进行 TRUNCATE 操作解码需要花费 4~5min，而在 PostgreSQL 14 中仅需要花费 1s。

- 支持两阶段提交解码到输出插件上。

逻辑复制支持两阶段提交，即对准备好的事务进行逻辑解码并将其发送到输出插件上，而不是在提交时进行解码。该特性允许两阶段分布式事务通过逻辑复制跨多个节点并且帮助降低节点回放延迟。PostgreSQL 14 订阅端的工作尚未完成，核心方案是使用该特性解码到输出插件上。

- 支持二进制形式传输数据。

在创建或修改订阅时，提供了一个 binary 选项，允许以二进制形式传输来自发布端的数据，以二进制形式传输的速度会更快。binary 选项的默认值为 false，打开 binary 选项后，只有具有二进制发送和接收函数的数据类型才会以二进制形式传输。

启用二进制形式的示例如下。

```
alter subscription mysub set (binary = on);
```

- 优化表的初始数据同步。

表的初始数据先由 tablesync worker 进程进行复制，再由 apply worker 进程进行同步。对于表的初始数据复制，在 PostgreSQL 14 之前的版本中使用临时复制槽在单个事务中完成，在进行表的初始数据复制时发生任何错误都会整体回滚，这对大表来说是非常"痛苦"的。基于此，逻辑复制在表的初始数据同步阶段进行了优化。在表的初始数据同步阶段允许有多个事务，如果在表的初始数据同步阶段发生错误，那么不再需要复制整个表，使用永久复制槽来跟踪 tablesync worker 进程的进度。
- 优化添加或删除发布。

在 PostgreSQL 14 之前的版本中,用户需要向订阅添加或删除发布时,只能使用 set publication 同时设置所有发布。如果一个订阅已经添加了两个发布,那么再次添加一个新发布时,需要同时设置三个发布。

```
alter subscription mysub set publication mypub1,mypub2,mypub3;
```

如果一个订阅已经包含很多发布,那么添加新发布是不太方便的。使用 ADD 命令或 DROP 命令可以对单个发布进行简化操作。

```
alter subscription mysub add publication mypub2;
```

- 添加逻辑复制监控视图。

PostgreSQL 14 专注于逻辑复制的改进,同时添加了 pg_stat_replication_slots 视图,用户可以了解逻辑复制槽的活动情况,可以监视流式传输到输出插件或订阅端,以及溢出到磁盘的数据量。

5.1.5 嵌套循环改进

PostgreSQL 14 新增了参数 enable_memoize,该参数默认打开,使用该参数可以对嵌套循环(Nested Loop)连接进行缓存记忆化来提高性能。下面通过示例进行演示。

创建 t1 表和 t2 表,每个表都有 10 万条数据,t1 表的 j 字段有 5 个不同的值,每个值重复 2 万次,t2 表的 j 字段有 2 万个不同的值,每个值重复 5 次。

```
create table t1 as
select i, i % 5 as j
from generate_series(1, 100000) as t(i);

create table t2 as
select i, i % 20000 as j
from generate_series(1, 100000) as t(i);

create index on t2(j);
```

观察下面语句的执行计划。

```
postgres=# explain select * from t1 join t2 on t1.j = t2.j;
                                  QUERY PLAN
-------------------------------------------------------------------------------
 Nested Loop  (cost=0.30..8945.41 rows=499500 width=16)
   ->  Seq Scan on t1  (cost=0.00..1443.00 rows=100000 width=8)
   ->  Memoize  (cost=0.30..0.41 rows=5 width=8)
         Cache Key: t1.j
         Cache Mode: logical
         ->  Index Scan using t2_j_idx on t2  (cost=0.29..0.40 rows=5 width=8)
               Index Cond: (j = t1.j)
(7 rows)
```

可以看到,在执行计划的 Memoize 节点对 t1 表的 j 字段进行缓存记忆的情况下,只需要进行 5 次查找,这是因为 t1 表的 j 字段只有 5 个不同的值。

如果显式关闭缓存记忆化,设置参数 enable_memoize 的值为 off,那么执行计划可能会发生改变,下面的示例使用了 HASH 连接。

```
postgres=# set enable_memoize to off;
SET
postgres=# explain select * from t1 join t2 on t1.j = t2.j;
                            QUERY PLAN
-----------------------------------------------------------------
 Hash Join  (cost=3084.00..11570.00 rows=499500 width=16)
   Hash Cond: (t1.j = t2.j)
   ->  Seq Scan on t1  (cost=0.00..1443.00 rows=100000 width=8)
   ->  Hash  (cost=1443.00..1443.00 rows=100000 width=8)
         ->  Seq Scan on t2  (cost=0.00..1443.00 rows=100000 width=8)
(5 rows)
```

5.1.6 并行特性增强

PostgreSQL 14 支持 RETURN QUERY 语句使用并行,并支持 postgres_fdw 模块使用并行。下面通过示例进行展示。

针对 RETURN QUERY 语句的并行特性,准备如下测试函数。

```
create function func_foo() returns setof t1
as $$
begin
    return query select * from t1 where id < 100;
end;
$$ language plpgsql;
```

通过 auto_explain 模块观察 func_foo 函数的执行计划。

```
set auto_explain.log_min_duration = '1ms';

select * from func_foo();
```

查看 RETURN QUERY 语句的执行计划可以发现,查询使用了两个并行工作进程。

```
2022-12-24 19:17:55 CST [15385] user=postgres,db=mydb, query_id=0 LOG:  statement:
    select * from func_foo();
2022-12-24 19:17:56 CST [15385] user=postgres,db=mydb, query_id=-
    7328796972449887879 LOG:  duration: 255.259 ms  plan:
Query Text: select * from t1 where id < 100
Gather  (cost=1000.00..10643.33 rows=100 width=36) (actual time=0.251..255.229
    rows=99 loops=1)
  Workers Planned: 2
  Workers Launched: 2
  Buffers: shared hit=4425
  ->  Parallel Seq Scan on t1  (cost=0.00..9633.33 rows=42 width=36) (actual
    time=141.552..226.517 rows=33 loops=3)
        Filter: (id < 100)
        Rows Removed by Filter: 333300
        Buffers: shared hit=4425
```

此外,PostgreSQL 14 增加了使用 postgres_fdw 模块查询远程数据库时的并行执行功能。当设置了 async_capable 标识时,可以使用异步并行查询,支持对 Foreign Scan 进行并行处理。

下面的示例中创建的分区表，使用远程分区，也就是分区表数据实际存储在远程服务器上。

```
create table local_part(c1 int, c2 varchar) partition by range(c1);

create foreign table remote_part_v1
    partition of local_part
for values from (0) to (1000)
server pgserver;

create foreign table remote_part_v2
    partition of local_part
for values from (1000) to (2000)
server pgserver;
```

在默认设置中，执行计划如下。

```
postgres=# explain(costs off) select * from local_part;
                QUERY PLAN
----------------------------------------------------------
 Append
   ->  Foreign Scan on remote_part_v1 local_part_1
   ->  Foreign Scan on remote_part_v2 local_part_2
(3 rows)
```

当在服务器上打开 async_capable 标识时，Foreign Scan 可以进行异步并行处理。

```
postgres=# alter server pgserver options (set async_capable 'on');
ALTER SERVER

postgres=# explain(costs off) select * from local_part;
                QUERY PLAN
----------------------------------------------------------
 Append
   ->  Async Foreign Scan on remote_part_v1 local_part_1
   ->  Async Foreign Scan on remote_part_v2 local_part_2
(3 rows)
```

除了可以在服务器上进行设置，也可以在外部表上进行设置。下面的示例中设置的效果是一样的。

```
alter server pgserver options (set async_capable 'off');

alter foreign table remote_part_v1 options (add async_capable 'on');
alter foreign table remote_part_v2 options (add async_capable 'on');
```

5.2 PostgreSQL 14 的可靠性提高

PostgreSQL 14 的可靠性提高主要体现在数据结构检测、备节点可作为恢复源、密码长度限制取消。

5.2.1 数据结构检测

对于数据结构的可靠性检测，PostgreSQL 14 进行了如下两个方面的强化。
- amcheck 模块增加了 verify_heapam 函数，实现了堆表数据逻辑错误检测功能，之前只能对 B-Tree 索引进行检测。
- 增加了 pg_amcheck 命令，可以对一个或多个数据库实施检测，可以选择要检测哪些模式、表和索引，要执行哪种检测，以及是否并行执行等。

5.2.2 备节点可作为恢复源

PostgreSQL 14 将 pg_rewind 工具在--source-server 选项中使用 standby 节点当作恢复源，示例如下。

初始化一个 1401 节点。
```
$ /opt/pg14/bin/initdb -D data1401
```
在 postgresql.conf 文件中设置如下参数。
```
port=1401
listen_addresses = '0.0.0.0'
wal_log_hints = on
wal_keep_size=100
```
这里将参数 wal_keep_size 的值设置得较大一些，以为 standby 节点保留足够多的 WAL 文件，这样可以增大 pg_rewind 工具同步恢复数据的概率。

启动 1401 节点。
```
$ /opt/pg14/bin/pg_ctl start -D data1401
```
搭建两个 standby 节点，即 1402 节点和 1403 节点，使用 pg_basebackup 工具拉取数据。
```
$ /opt/pg14/bin/pg_basebackup -D data1402 -h 127.0.0.1 -p1401
$ /opt/pg14/bin/pg_basebackup -D data1403 -h 127.0.0.1 -p1401
```
分别修改 1402 节点和 1403 节点的 postgresql.conf 文件，修改参数 port 和 primary_conninfo 的值。1402 节点的 postgresql.conf 文件修改如下。
```
port=1402
primary_conninfo = 'host=127.0.0.1 port=1401 user=postgres application_name=1402'
```
1403 节点的 postgresql.conf 文件修改如下。
```
port=1403
primary_conninfo = 'host=127.0.0.1 port=1401 user=postgres application_name=1403'
```
分别创建 1402 节点和 1403 节点的 standby 模式触发文件并启动服务器。1402 节点的操作如下。
```
$ touch data1402/standby.signal
$ /opt/pg14/bin/pg_ctl start -D data1402
```
1403 节点的操作如下。
```
$ touch data1403/standby.signal
$ /opt/pg14/bin/pg_ctl start -D data1403
```
至此，主备环境搭建完成。下面从 1401 节点上查看流复制状态。
```
$ /opt/pg14/bin/psql -p1401
```

```
postgres=# select usename,application_name,client_addr,
        client_port,state,sync_state from pg_stat_replication;
 usename | application_name | client_addr | client_port | state   |sync_state
---------+------------------+-------------+-------------+---------+---------
 postgres| 1402             | 127.0.0.1   |       34378 |streaming| async
 postgres| 1403             | 127.0.0.1   |       34380 |streaming| async
(2 rows)
```

下面模拟 1401 节点和 1402 节点形成双主库，对 1402 节点执行 promote 操作，在执行该操作之前不关闭主库的 1401 节点。

```
$ /opt/pg14/bin/psql -p1402
select pg_promote();
```

1402 节点变为新主库之后，先恢复旧主库 1401 节点，使用 pg_rewind 工具恢复之前需要关闭 1401 节点。

```
$ /opt/pg14/bin/pg_ctl stop -D data1401
```

使用 pg_rewind 工具对 1401 节点同步数据。

```
$ /opt/pg14/bin/pg_rewind -D data1401 --source-server="host=127.0.0.1 port=1402
    user=postgres"
pg_rewind: servers diverged at WAL location 0/4000060 on timeline 1
pg_rewind: rewinding from last common checkpoint at 0/3000060 on timeline 1
pg_rewind: Done!
```

同步数据完成之后，修改 1401 节点的 postgresql.conf 文件，参数 primary_conninfo 修改如下。

```
primary_conninfo = 'host=127.0.0.1 port=1402 user=postgres application_name=1401'
```

创建 1401 节点的 standby 模式触发文件并启动服务器。

```
$ touch data1401/standby.signal
$ /opt/pg14/bin/pg_ctl start -D data1401
```

在 1402 节点上查看流复制状态。

```
$ /opt/pg14/bin/psql -p1402
postgres=# select usename,application_name,client_addr,
        client_port,state,sync_state from pg_stat_replication;
 usename | application_name | client_addr | client_port | state   |sync_state
---------+------------------+-------------+-------------+---------+--------
 postgres| 1401             | 127.0.0.1   |       40976 |streaming| async
(1 row)
```

可以看到，1402 节点与 1401 节点已经建立了主备关系。

此时，1401 节点已切换为备库，1403 节点与 1401 节点的流复制进程连接失败，1403 节点服务被关闭。

恢复 1403 节点的流复制连接可以从 1401 节点发起。使用 pg_rewind 工具连接 1401 节点进行数据同步的操作如下。

```
$ /opt/pg14/bin/pg_rewind -D data1403 --source-server="host=127.0.0.1 port=1401
    user=postgres"
pg_rewind: servers diverged at WAL location 0/4000060 on timeline 1
pg_rewind: rewinding from last common checkpoint at 0/3000060 on timeline 1
pg_rewind: Done!
```

如果在 PostgreSQL 14 之前的版本中连接 1401 节点进行数据同步，那么会有如下错误提示。
```
pg_rewind: fatal: source server must not be in recovery mode
```
在 1402 节点上查看流复制状态。
```
$ /opt/pg14/bin/psql -p1402
postgres=# select usename,application_name,client_addr,
        client_port,state,sync_state from pg_stat_replication;
 usename  | application_name | client_addr | client_port |  state   |sync_state
----------+------------------+-------------+-------------+----------+--------
 postgres | 1401             | 127.0.0.1   |       40976 |streaming| async
 postgres | 1403             | 127.0.0.1   |       41000 |streaming| async
(2 rows)
```
可以看到，1402 节点与 1401 节点、1403 节点的流复制状态正常。

5.2.3 密码长度限制取消

在 PostgreSQL 14 之前的版本中在使用 SCRAM-SHA-256 方式对口令进行加密时，用户的密码长度是有限制的。例如，从下面的示例中可以看到，密码长度不能超过 1024 个字符。

```
postgres=# set password_encryption to 'scram-sha-256';
SET

postgres=# select 'drop user if exists testuser; create user testuser password '||
   quote_literal(repeat('a',1024))||';' \gexec
CREATE ROLE

postgres=# select 'drop user if exists testuser; create user testuser password '||
   quote_literal(repeat('a',1025))||';' \gexec
ERROR:  password too long
```

从 PostgreSQL 14 开始，在使用 SCRAM-SHA-256 方式对口令进行加密时取消了这个限制。下面是在 PostgreSQL 14 中使用 10 万个字符进行测试的示例，很明显可以创建成功。

```
postgres=# show password_encryption;
 password_encryption
---------------------
 scram-sha-256
(1 row)

postgres=# select 'drop user if exists testuser; create user testuser password '||
   quote_literal(repeat('a',100000))||';' \gexec
NOTICE:  role "testuser" does not exist, skipping
CREATE ROLE
```

5.3　PostgreSQL 14 的运维管理优化

PostgreSQL 14 的运维管理优化主要体现在查询 ID 引入、索引表空间在线移动、触发器在线重建、控制客户端连接、后台操作进度报告引入、可观测性增强。

5.3.1 查询 ID 引入

有 Oracle 运维背景的 DBA 在进行 SQL 语句优化及问题排查时，不可避免地会用到 SQL_ID，在 PostgreSQL 中与 SQL_ID 概念对应的是查询 ID，即查询语句的唯一标识。

在 PostgreSQL 14 之前的版本中需要通过 pg_stat_statements 模块生成查询 ID，PostgreSQL 14 引入了一个参数 compute_query_id，可以控制内部直接计算生成。同时，活动会话视图也增加了查询 ID 的 id 字段，其他依赖查询 ID 的扩展插件可以直接使用计算的查询 ID 来提升性能。

下面演示查询 ID 使用的过程。

对 postgresql.conf 文件进行设置。

```
shared_preload_libraries = 'pg_stat_statements'
compute_query_id = on
log_duration = on
log_statement = 'all'
log_line_prefix = '%t [%p] user=%u,db=%d,app=%a,client=%h,query_id=%Q '
```

在上述代码中，前两行加载了 pg_stat_statements 模块，并显式打开了生成查询 ID 的开关，以便明确使用这个特性而没有歧义，最后一行对参数 log_line_prefix 进行了定制化配置，%Q 表示可以在日志中查看查询 ID。

重启数据库后，新建 pgbench 数据库，并创建 pg_stat_statements 模块。

```
createdb pgbench
psql -d pgbench -c "create extension pg_stat_statements;"
```

对新建的 pgbench 数据库使用 pgbench 工具进行初始化并模拟一些查询。

```
pgbench -i -s 10 pgbench
pgbench -c 5 -j 1 -s 10 -T 60 pgbench
```

以下代码用于返回 pg_stat_statements 视图 5 个经常运行的查询。

```
select queryid, query, calls, total_exec_time
  from pg_stat_statements
 where query not in ('begin', 'end')
order by calls desc
limit 5;
```

从下面的查询结果中可以看出，每个查询都运行了 50067 次，查询中的第一列查询 ID 后面是参数化的查询文本。

```
-[ RECORD 1 ]---+-------------------------------------------------------
queryid         | 7813238744472522214
query           | INSERT INTO pgbench_history (tid, bid, aid, delta, mtime) VALUES
   ($1, $2, $3, $4, CURRENT_TIMESTAMP)
calls           | 50067
total_exec_time | 779.829668000005
-[ RECORD 2 ]---+-------------------------------------------------------
queryid         | 617649163336858051
query           | UPDATE pgbench_tellers SET tbalance = tbalance + $1 WHERE tid = $2
calls           | 50067
```

```
total_exec_time    | 7488.97469200003
-[ RECORD 3 ]---+------------------------------------------------
queryid            | -1707234916255296174
query              | UPDATE pgbench_accounts SET abalance = abalance + $1 WHERE aid = $2
calls              | 50067
total_exec_time    | 2900.1079170000025
-[ RECORD 4 ]---+------------------------------------------------
queryid            | 7229793815732221840
query              | SELECT abalance FROM pgbench_accounts WHERE aid = $1
calls              | 50067
total_exec_time    | 902.3938220000007
-[ RECORD 5 ]---+------------------------------------------------
queryid            | 286400176462488288
query              | UPDATE pgbench_branches SET bbalance = bbalance + $1 WHERE bid = $2
calls              | 50067
total_exec_time    | 35189.59980300006
```

下面模拟实际环境中执行时间较长的慢查询，设置一个人为的慢查询，选择上面最后一个针对 pgbench_branches 表的 UPDATE 操作，查询 ID 为 286400176462488288，这个慢查询的执行时间是 5 个查询中最短的。

手动执行下面的语句。

```
update public.pgbench_branches
   set bbalance = bbalance + 1
 where bid = 1;
```

执行一次 UPDATE 操作之后，检查 pg_stat_statements 视图，calls 列的值由 50067 变为了 50068。

```
pgbench=# select queryid, calls
            from pg_stat_statements
           where queryid = 286400176462488288;
      queryid       | calls
--------------------+-------
 286400176462488288 | 50068
(1 row)
```

为了模拟慢查询并观察效果，使用 3 个会话连接，第一个会话中显式启动事务并更新 bid=1，不提交事务（在生产环境中不应该这样做）。

```
pgbench=# start transaction;
START TRANSACTION

pgbench=*# update public.pgbench_branches
             set bbalance = bbalance + 1
           where bid = 1;
UPDATE 1
```

第二个会话中在 bid=1 上执行相同的 UPDATE 操作，这次不使用 START TRANSACTION，此时语句因第一个会话显式启动事务持有锁而被挂起。

当第二个 UPDATE 操作运行并处于等待中时，打开第三个会话，使用实时查询视图和历史查询视图（pg_stat_activity 视图和 pg_stat_statements 视图）进行连接查询，可以对当前运行的查询与历史查询的统计信息进行比较。

```
select a.pid, a.query_id,
    now() - a.query_start AS query_duration,
    a.state,
    now() - a.state_change AS state_duration,
    a.wait_event,
    ss.calls,
    ROUND(ss.total_exec_time) AS total_exec_time,
    ROUND(ss.mean_exec_time) AS mean_exec_time,
    ROUND(ss.stddev_exec_time) AS stddev_exec_time
  from pg_catalog.pg_stat_activity a
inner join pg_stat_statements ss on a.query_id = ss.queryid
where query_id = 286400176462488288;
```

查看 state 列的值为 active 的记录如下。

```
-[ RECORD 2 ]----+--------------------
pid              | 9840
query_id         | 286400176462488288
query_duration   | 00:00:07.382018
state            | active
state_duration   | 00:00:07.382013
wait_event       | transactionid
calls            | 50069
total_exec_time  | 35190
mean_exec_time   | 1
stddev_exec_time | 5
```

查看 query_duration 列的值，并将其与 mean_exec_time 列的值和 stddev_exec_time 列的值进行比较。mean_exec_time 列的值和 stddev_exec_time 列的值的单位均为毫秒（ms）。平均执行时间仅为 1ms，标准偏差为 5ms，当前运行的查询时间为 7.382013s，表示与正常性能预期有明显偏差。该查询结果有助于确定当前性能与历史性能的对比情况。

在使用 EXPLAIN 命令结合 verbose 选项查看执行计划时也会显示查询 ID，这样当请求他人帮助分析复杂查询的性能时会比较方便。

```
pgbench=# explain (verbose, costs off)
         update public.pgbench_branches
             set bbalance = bbalance + 1
         where bid = 1;
              QUERY PLAN
------------------------------------------
Update on public.pgbench_branches
  -> Seq Scan on public.pgbench_branches
       Output: (bbalance + 1), ctid
       Filter: (pgbench_branches.bid = 1)
Query Identifier: 286400176462488288
(5 rows)
```

使用了前面配置的参数 log_duration 和 log_line_prefix 后，日志文件中将输出以下内容。

```
2022-12-04 11:11:28 CST [4881] user=postgres,db=pgbench,app=pgbench,
client=[local],query_id=7229793815732221840 LOG:  duration: 0.144 ms
```

可以将某个查询 ID 作为关键字使用下面的语句来查询相关的日志文件。

```
$ cat /opt/pgdata1406/log/pg_log_7.log \
```

```
        | grep -m 5 query_id=286400176462488288
2022-12-04 11:11:27 CST [4885] user=postgres,db=pgbench,app=pgbench,
client=[local],query_id=286400176462488288 LOG:  duration: 0.328 ms
2022-12-04 11:11:27 CST [4882] user=postgres,db=pgbench,app=pgbench,
client=[local],query_id=286400176462488288 LOG:  duration: 0.299 ms
2022-12-04 11:11:27 CST [4884] user=postgres,db=pgbench,app=pgbench,
client=[local],query_id=286400176462488288 LOG:  duration: 0.303 ms
2022-12-04 11:11:27 CST [4883] user=postgres,db=pgbench,app=pgbench,
client=[local],query_id=286400176462488288 LOG:  duration: 0.298 ms
2022-12-04 11:11:27 CST [4881] user=postgres,db=pgbench,app=pgbench,
client=[local],query_id=286400176462488288 LOG:  duration: 0.519 ms
```

使用查询 ID 是 PostgreSQL 14 中很重要的一个后端改进，pg_stat_activity 视图、EXPLAIN 命令、pg_stat_statements 视图都可以共享一个查询 ID，同时也可以在日志文件中查看。

5.3.2 索引表空间在线移动

在 PostgreSQL 14 中，重建索引操作新增了一个 TABLESPACE 选项，在重建索引时支持将索引移动到新表空间中。TABLESPACE 选项可以与 CONCURRENTLY 选项同时使用，进行重建索引操作时，在当前索引及其父表上使用 Share Update Exclusive 锁允许重建索引时并发读取和写入，并移动到新磁盘分区。

新增的 TABLESPACE 选项有以下几个限制。

- 对于分区表，leaf 分区的索引会移动到新表空间中，非 leaf 对象在 pg_class 表的 reltablespace 列仍然记录为原表空间。
- TOAST 表的索引保留在其原表空间中。
- 系统元数据不允许执行该操作，系统表的索引不能移动到新表空间中。

下面通过分区表的一个示例进行演示。

```
create table parent_tab (id int) partition by range (id);
create index parent_index on parent_tab (id);
create table child_0_10 partition of parent_tab
   for values from (0) to (10);
create table child_10_20 partition of parent_tab
   for values from (10) to (20);
```

创建了一个简单的分区树，包含一个名为 parent_tab 的分区表和两个 leaf 分区。

```
postgres=# select * from pg_partition_tree('parent_tab');
   relid     | parentrelid | isleaf | level
-------------+-------------+--------+-------
 parent_tab  |             | f      |   0
 child_0_10  | parent_tab  | t      |   1
 child_10_20 | parent_tab  | t      |   1
(3 rows)
```

分区表的索引如下。

```
postgres=# select * from pg_partition_tree('parent_index');
     relid          | parentrelid | isleaf | level
--------------------+-------------+--------+-------
```

```
 parent_index       |                    | f    |    0
 child_0_10_id_idx  | parent_index       | t    |    1
 child_10_20_id_idx | parent_index       | t    |    1
(3 rows)
```

下面对分区表使用 reindex(tablespace)语句。

```
postgres=# \db extra_tbspace
         List of tablespaces
    Name      |  Owner   |    Location
--------------+----------+------------------
 extra_tbspace | postgres | /opt/mytalespace
(1 row)
postgres=# reindex (concurrently,tablespace 'extra_tbspace') table parent_tab;
REINDEX
```

执行如下代码可以发现，所有索引都移动到了新表空间中。

```
postgres=# select c.relname, t.spcname
        from pg_partition_tree('parent_index') p
        join pg_class c on (c.oid = p.relid)
        join pg_tablespace t on (c.reltablespace = t.oid);
       relname       |    spcname
--------------------+---------------
 child_0_10_id_idx  | extra_tbspace
 child_10_20_id_idx | extra_tbspace
(2 rows)
```

执行如下代码可以发现，分区表的索引 parent_index 的表空间引用没有发生变化。

```
postgres=# select reltablespace from pg_class where relname='parent_index';
 reltablespace
---------------
             0
(1 row)
```

REINDEX(CONCURRENTLY)移动表空间相比 ALTER TABLE 操作看起来更复杂，但使用较低的锁。如果在 REINDEX 之后未使用 ALTER TABLE ONLY 更改表空间，则任何新分区仍将使用父表的表空间，这一点需要注意。

5.3.3 触发器在线重建

PostgreSQL 14 之前的版本仅支持执行 CREATE TRIGGER 命令，它定义了一个触发器，该触发器将在相关事件发生时运行指定的函数。在 PostgreSQL 14 中支持执行 CREATE OR REPLACE TRIGGER 命令，该命令可以用于创建新触发器或覆盖现有触发器，以减少从 Oracle 中迁移到 PostgreSQL 14 中所需的应用程序迁移工作。

对于 PostgreSQL 14 之前的版本，如果要修改触发器，那么需要先删除再创建，此时会有一个小的窗口期不能使用触发器，通常可以在一个事务中进行操作，示例如下。

```
begin;
drop trigger trg1 on my_table;
create trigger trg1 before insert on my_table for each row execute procedure
    my_function();
end;
```

PostgreSQL 14 可以直接执行 CREATE OR REPLACE TRIGGER 命令在单个语句中替换现有触发器，并使用侵入性较小的 Share Row Exclusive 锁。

```
begin;
create or replace trigger trg1 before insert on my_table for each row execute
    procedure my_function();
end;
```

在替换触发器时，也可以修改触发器的级别，如将行级触发器改为语句级触发器。

```
create or replace trigger trg1 before insert on my_table for each statement execute
    procedure my_function();
```

5.3.4 控制客户端连接

在 PostgreSQL 14 中，可以设置如下参数。
- idle_in_transaction_session_timeout。
- idle_session_timeout。
- client_connection_check_interval。
- lock_timeout。
- statement_timeout。

如果应用程序经常出现锁等待的情况，那么需要设置参数 lock_timeout 来控制超时时间。参数 statement_timeout 用于控制任意语句执行的超时时间，通常很少限制语句的执行时间。如果连接活动会话视图可以观测到 idle in transaction 状态的连接，那么应该设置参数 idle_in_transaction_session_timeout，根据实际业务场景设置 5～30min 不等。否则，该连接的事务一直保持打开状态，会引起 VACUUM 操作问题，并可能因表膨胀而造成性能下降和产生糟糕的用户体验。

下面演示参数 idle_in_transaction_session_timeout 的特性。

```
postgres=# set idle_in_transaction_session_timeout to '3s';
SET
postgres=# begin;
BEGIN
postgres=# --wait 3 seconds
postgres=# select 1;
FATAL:  terminating connection due to idle-in-transaction timeout
```

上面的示例中设置等待 3s，一旦事务空闲时间过长，连接就会被服务器自动中断，通过调整参数 idle_in_transaction_session_timeout 可以轻松地防止事务空闲时间过长带来的恶劣影响。

idle_session_timeout 和 client_connection_check_interval 是 PostgreSQL 14 新增的参数。参数 client_connection_check_interval 可以检测运行中的语句是否已经失去连接，如果检测到客户端连接已离线，那么可以快速中断运行中的语句。

参数 idle_session_timeout 的功能与参数 idle_in_transaction_session_timeout 的功能类似，如果空闲连接比较多，那么会占用大量的内存，参数 idle_session_timeout 用于自动杀死那些空闲时间超过设置值且不在事务中的空闲连接，可以释放缓存，避免出现 OOM 等问题。

下面演示参数 idle_session_timeout 的特性。新建会话连接，设置超时时间为 5s。

```
postgres=# set idle_session_timeout to '5s';
SET
postgres=# --wait 5 seconds
postgres=# select 1;
FATAL:  terminating connection due to idle-session timeout
```

超过 5s 后进行操作，可以看到会话连接被系统自动释放。

5.3.5 后台操作进度报告引入

PostgreSQL 14 新引入了一个后台操作进度报告，即 pg_stat_progress_copy 视图，用于跟踪 COPY 操作执行的进度。该视图常用于在逻辑复制场景中监控发布端和订阅端的表初始数据同步的进度。

执行下面的代码，每隔 1 秒进行观察输出。

```
select datname,relid::regclass as table,
       command,type,
       bytes_processed,tuples_processed
  from pg_stat_progress_copy \watch 1
```

在发布端可以观测到通过 COPY 操作发送的字节量及行数。

```
               Wed 23 Feb 2022 07:01:46 AM CST (every 1s)
 datname  | table | command | type  | bytes_processed | tuples_processed
----------+-------+---------+-------+-----------------+------------------
 postgres | test  | COPY TO | PIPE  |       932960052 |          9540522
(1 row)
```

在订阅端可以观测到通过 COPY 操作接收的字节量及行数。

```
               Wed 23 Feb 2022 07:01:47 AM CST (every 1s)
 datname  | table | command   | type     | bytes_processed | tuples_processed
----------+-------+-----------+----------+-----------------+------------------
 postgres | test  | COPY FROM | CALLBACK |       948074690 |          9694752
(1 row)
```

目前，PostgreSQL 14 及 PostgreSQL 14 之前的版本支持的后台操作进度报告如表 5-1 所示。

表 5-1　PostgreSQL 14 及 PostgreSQL 14 之前的版本支持的后台操作进度报告

进度报告名称	描述	版本
pg_stat_progress_copy	跟踪 COPY 操作执行的进度	PostgreSQL 14
pg_stat_progress_basebackup	实时跟踪使用 pg_basebackup 工具备份的各个阶段	PostgreSQL 13
pg_stat_progress_analyze	ANALYZE 操作执行的进度	PostgreSQL 13
pg_stat_progress_create_index	在创建或重建索引时，显示当前创建索引的各个阶段的信息	PostgreSQL 12
pg_stat_progress_cluster	显示当前操作正在执行的各个阶段的信息	PostgreSQL 12
pg_stat_progress_vacuum	显示清理操作正在执行的各个阶段的信息	PostgreSQL 9.6

5.3.6 可观测性增强

PostgreSQL 14 增加了一些可观测性视图，包括 pg_backend_memory_contexts 视图、pg_stat_progress_copy 视图、pg_stat_replication_slots 视图、pg_stat_wal 视图、pg_stats_ext_exprs 视图。

当某个客户端连接占用较多内存时，可以使用新增的 pg_backend_memory_contexts 视图详细查看当前会话的内存上下文使用情况。而 pg_log_backend_memory_contexts 函数可以使用输出到数据库日志文件的形式来记录指定进程的内存消耗信息。

为了更好地监控逻辑复制及逻辑复制槽的活动状态，可以使用新增的 pg_stat_progress_copy 视图观测发布端和订阅端初始数据同步的进度，使用新增的 pg_stat_replication_slots 视图查看全局的逻辑复制槽当前状态信息，特别是有关大事务溢出到磁盘中的统计信息。针对 pg_stat_replication_slots 视图的统计信息，可以使用 pg_stat_reset_replication_slot(slot_name)函数来重置单个指定逻辑复制槽的统计信息，也可以使用 pg_stat_reset_replication_slot(NULL)函数来重置所有逻辑复制槽的统计信息。

WAL 的活动统计信息可以通过新增的 pg_stat_wal 视图来跟踪。打开可选的参数 track_wal_io_timing 可以获取与 WAL 相关的 I/O 详细信息，可以查询到 WAL 写入和 WAL 同步到磁盘的 I/O 时间。不过，该设置可能会产生明显的开销，默认关闭。

新增的 pg_stats_ext_exprs 视图用于存储扩展统计信息对象中包含的表达式信息，以让用户直观地看到扩展统计信息中的表达式信息，便于日常维护和优化分析。

5.4 PostgreSQL 14 的开发易用性提升

PostgreSQL 14 的开发易用性提升主要体现在多范围类型引入、存储过程支持 OUT 模式参数、新形式的 SQL 函数引入、JSON 操作功能增强、递归查询改进、易用的内置函数引入。

5.4.1 多范围类型引入

PostgreSQL 14 之前的版本支持以下几种标准的范围类型。
- int4range、int8range：int 和 bigint 范围类型。
- numrange：numeric 范围类型。
- tstzrange、daterange、tsrange：timestamptz、date 和 time 范围类型。

PostgreSQL 14 新引入了多范围类型，多范围类型是一组不重叠的范围类型，可以有效地对具有间隙的范围进行建模。例如，医生跟踪病人住了多少天院，可以将其存储为 datemultirange 类型。

多范围类型与现有的范围类型类似：
- int4multirange、int8multirange：int 和 bigint 多范围类型。

- nummultirange：numeric 多范围类型。
- tstzmultirange、datemultirange、tsmultirange：timestamptz、date 和 time 多范围类型。

多范围类型的规范形式由一对花括号后跟逗号分隔的范围列表组成，所有边界都以 "[" 开始，以 ")" 结束，示例如下。

```
postgres=# SELECT '{[2021-09-02, 2021-09-05),
                    [2021-09-08, 2021-09-12)}'::datemultirange;
              datemultirange
---------------------------------------------------
 {[2021-09-02,2021-09-05),[2021-09-08,2021-09-12)}
(1 row)
```

如果边界有重叠或包含关系，那么范围会自动合并。

```
postgres=# SELECT '{[2022-06-01, 2022-07-02),
                    [2022-09-03, 2022-09-12),
                    [2022-09-01, 2022-09-10)}'::datemultirange;
              datemultirange
---------------------------------------------------
 {[2022-06-01,2022-07-02),[2022-09-01,2022-09-12)}
(1 row)
```

从以上代码中可以看到，后面的两个日期范围合并成了一个日期范围。

5.4.2 存储过程支持 OUT 模式参数

PostgreSQL 11 开始支持存储过程，与函数相比，存储过程允许用户进行灵活的事务控制，可以手动开启事务、设置保存点、提交和回滚事务。另外，存储过程不强制设置返回值，如果只需要执行业务语句，不需要返回客户端结果，那么在服务端执行的存储过程效率会比较高。如果需要返回一个或多个标量值，那么在 PostgreSQL 14 之前的版本中只能使用 INOUT 模式参数，使用 OUT 模式参数会出现如下错误提示。

```
postgres=# CREATE PROCEDURE p_sum(IN p1 bigint,IN p2 bigint,OUT p3 bigint)
           AS $$
           BEGIN
              p3 := p1 + p2 ;
           END;
           $$ LANGUAGE plpgsql ;
ERROR:  procedures cannot have OUT arguments
HINT:  INOUT arguments are permitted.
```

从 PostgreSQL 14 开始，支持在存储过程中直接使用 OUT 模式参数。在 PostgreSQL 14 中使用上面的代码可以执行成功。在存储过程中直接使用 OUT 模式参数，可以使从其他数据库进行的迁移工作变得更加轻松。

5.4.3 新形式的 SQL 函数引入

在声明函数体时，需要使用 "$$" 来标记代码为字符串常量。例如，两个整数相加的一个简单函数的示例如下。

```
create function f_add(int, int)
```

```
returns int
language sql
as $$
  select $1 + $2;
$$;
```

要编写另外一个函数，可以调用上面定义的函数。例如，使用 **f_add** 函数编写对多个整数相加的 **f_add_nest** 函数的示例如下。

```
create function f_add_nest(int, int, int, int)
returns int
language sql
as $$
  select f_add(f_add($1, $2), f_add($3, $4));
$$;
```

创建上述两个函数后，运行 **f_add_nest** 函数，返回如下结果。

```
postgres=# select f_add_nest(1,1,2,2);
 f_add_nest
------------
          6
(1 row)
```

下面删除 **f_add** 函数。

```
postgres=# drop function f_add(int, int);
DROP FUNCTION
```

删除 **f_add** 函数之后，再次运行 **f_add_nest** 函数会失败。

```
postgres=# select f_add_nest(1,1,2,2);
ERROR:  function f_add(integer, integer) does not exist
LINE 2:    select f_add(f_add($1, $2), f_add($3, $4));
                  ^
HINT:  No function matches the given name and argument types. You might need to add
   explicit type casts.
QUERY:
  select f_add(f_add($1, $2), f_add($3, $4));

CONTEXT:  SQL function "f_add_nest" during inlining
```

在 PostgreSQL 14 之前的版本中创建的自定义函数缺少依赖性检测，删除某个函数后可能会破坏其他函数，使用 PostgreSQL 14 之前的版本无法跟踪函数主体内的依赖关系。

PostgreSQL 14 对使用 SQL 函数或存储过程增加了一种新语法形式，函数体使用 BEGIN ATOMIC 语句，在声明函数体时，不再需要使用 "$$" 来标记代码为字符串常量。

在 PostgreSQL 14 中再次使用上面的示例。

```
create function f_add2(int, int)
returns int
language sql
begin atomic;
  select $1 + $2;
end;
```

上述代码中不再使用 "$$" 来标记代码为字符串常量，而使用实际的代码，可以通过新增的 **pg_get_function_sqlbody** 函数来查看函数的定义。

```
postgres=# select pg_get_function_sqlbody('f_add2'::regproc);
```

```
 pg_get_function_sqlbody
-------------------------
 BEGIN ATOMIC            +
  SELECT ($1 + $2);      +
 END
(1 row)
```

创建 f_add_nest2 函数，f_add_nest2 函数调用 f_add2 函数。

```
create function f_add_nest2(int, int, int, int)
returns int
language sql
begin atomic;
  select f_add2(f_add2($1, $2), f_add2($3, $4));
end;
```

与预期的一样，f_add_nest2 函数正常工作。

```
postgres=# select f_add_nest2(1,1,2,2);
 f_add_nest2
-------------
           6
(1 row)
```

下面尝试删除 f_add2 函数。

```
postgres=# drop function f_add2(int, int);
ERROR:  cannot drop function f_add2(integer,integer) because other objects depend
  on it
DETAIL:  function f_add_nest2(integer,integer,integer,integer) depends on function
  f_add2(integer,integer)
HINT:  Use DROP ... CASCADE to drop the dependent objects too.
```

从上述代码中可以看出，使用新形式的 SQL 函数可以防止用户删除某个函数，避免对有依赖引用的函数产生损坏。

如果使用 CASCADE 选项进行删除，那么会删除所有依赖的函数。

```
postgres=# drop function f_add2(int, int) cascade;
NOTICE:  drop cascades to function f_add_nest2(integer,integer,integer,integer)
DROP FUNCTION
```

使用 CASCADE 选项清理测试函数比较方便，但在生产环境中进行级联删除时要小心，以免意外删除重要对象。

新形式的 SQL 函数可以帮用户更好地管理自定义函数，防止用户意外删除函数，目前推荐使用这个特性。

5.4.4　JSON 操作功能增强

JSON 数据类型在 PostgreSQL 14 中有两种格式：JSON 和 JSONB。JSON 是文本格式，在插入时比较高效，如果追求快速入库，那么采用 JSON 存储会比较合适。JSONB 与 JSON 不同，JSONB 是二进制形式，在存储时会删除空格及重复的键值对，且所有键值对已排好序。在生产环境中，主要使用 JSONB，这样能够更好地使用索引。

JSON 有两种基本的运算操作符，一种用于返回文本数据，另一种用于返回 JSON 数据。

```
postgres=# select pg_typeof(details ->> 'attributes') from shirts where id=1;
```

```
 pg_typeof
-----------
 text
(1 row)

postgres=# select pg_typeof(details -> 'attributes') from shirts where id=1;
 pg_typeof
-----------
 jsonb
(1 row)
```

在 PostgreSQL 14 之前的版本中,根据衬衫的颜色和尺码查询的代码如下。

```
select *
 from shirts
where details->'attributes'->>'color' = 'yellow'
  and details->'attributes'->>'size' = 'medium';
```

在 PostgreSQL 14 中采用数组下标的形式访问 JSON 的代码如下。

```
select *
 from shirts
where details['attributes']['color'] = '"yellow"'
  and details['attributes']['size'] = '"medium"';
```

注意,等号右侧需要采用 JSON 的字符串形式,用双引号引起来。

更新记录的代码如下。

```
update shirts
   set details['attributes']['color'] = '"blue"'
 where id = 1;
```

5.4.5 递归查询改进

PostgreSQL 14 针对递归查询增加了 SEARCH 选项和 CYCLE 选项。这两个选项的使用极大地简化了编写递归查询的方法。下面通过一个旅行计划的示例进行展示。

假设想在武汉、长沙、南昌几个城市之间穿梭旅行,并且有预算。

创建城市表。

```
create table cities(
city_id int primary key,
city_name varchar
);

insert into cities values (0, '武汉'),
                          (1, '长沙'),
                          (2, '南昌');
```

通过 App、公众号或网站获取在武汉、长沙、南昌几个城市之间穿梭旅行的费用。

```
create table trips(
trip_id int primary key,
city_from int references cities(city_id),
city_to int references cities(city_id),
price int
);
```

```
insert into trips values
    (1, 0, 1, 292),
    (2, 0, 2, 253),
    (3, 1, 0, 292),
    (4, 1, 2, 287),
    (5, 2, 0, 253),
    (6, 2, 1, 287);
```

在上面创建的 trips 表中，city_from 表示出发地，city_to 表示目的地，price 表示旅行费用。trips 表中有所有可用路线，以及相关成本，如 trip_id 的值为 2 的路线，从武汉出发到达南昌需要花费 253 元。

当选择某个城市作为起点时，通过 trips 表关联 cities 表能够确定目的地。例如，从武汉出发的代码如下。

```
select src.city_name,
       dst.city_name,
       trips.price
  from cities src
  join trips on src.city_id = trips.city_from
  join cities dst on trips.city_to = dst.city_id
 where src.city_name='武汉';
```

目的地可以为长沙也可以为南昌。

```
city_name  | city_name | price
-----------+-----------+-------
武汉       | 长沙      | 292
武汉       | 南昌      | 253
(2 rows)
```

使用递归查询可以遍历所有可能的组合。为了避免递归查询陷入无限循环，假设总预算为 1000 元。递归查询的代码如下。

```
with recursive trip_journey(
    city_id,
    trip_id,
    total_price,
    journey_stops
) as (
    select
        city_id as city_id,
        null::int as trip_id,
        0 price_in,
        array[city_name] as journey_name
    from cities
    where city_id=0
    union all
    select
        trips.city_to,
        trips.trip_id,
        tj.total_price + trips.price,
        tj.journey_stops || city_b.city_name
    from trip_journey tj join trips on tj.city_id = trips.city_from
```

```
    join cities city_a on trips.city_from = city_a.city_id
    join cities city_b on trips.city_to = city_b.city_id
    where tj.total_price + trips.price < 1000
) select * from trip_journey;
```

下面为递归查询的结果。

```
 city_id | trip_id | total_price |     journey_stops
---------+---------+-------------+----------------------
       0 |         |           0 | {武汉}
       1 |       1 |         292 | {武汉,长沙}
       2 |       2 |         253 | {武汉,南昌}
       0 |       3 |         584 | {武汉,长沙,武汉}
       2 |       4 |         579 | {武汉,长沙,南昌}
       0 |       5 |         506 | {武汉,南昌,武汉}
       1 |       6 |         540 | {武汉,南昌,长沙}
       1 |       1 |         798 | {武汉,南昌,武汉,长沙}
       1 |       1 |         876 | {武汉,长沙,武汉,长沙}
       2 |       2 |         759 | {武汉,南昌,武汉,南昌}
       2 |       2 |         837 | {武汉,长沙,武汉,南昌}
       0 |       3 |         832 | {武汉,南昌,长沙,武汉}
       2 |       4 |         827 | {武汉,南昌,长沙,南昌}
       0 |       5 |         832 | {武汉,长沙,南昌,武汉}
       1 |       6 |         866 | {武汉,长沙,南昌,长沙}
(15 rows)
```

上面的代码中总结了所有可能的行程。那么如何排序呢？PostgreSQL 14 提供了两种算法。

- Breadth-First Search（BFS）：广度优先查找。
- Depth-First Search（DFS）：深度优先查找。

如果按照广度优先查找的算法进行排序，使用下面的语句替换递归查询最后的语句。

```
search breadth first by city_id set ordercol
select * from trip_journey order by ordercol limit 10;
```

那么结果集按 ordercol 列的第一个数字（行程落脚点的数量）进行排序。

```
 city_id | trip_id | total_price |   journey_stops    | ordercol
---------+---------+-------------+--------------------+----------
       0 |         |           0 | {武汉}             | (0,0)
       1 |       1 |         292 | {武汉,长沙}        | (1,1)
       2 |       2 |         253 | {武汉,南昌}        | (1,2)
       0 |       3 |         584 | {武汉,长沙,武汉}   | (2,0)
       0 |       5 |         506 | {武汉,南昌,武汉}   | (2,0)
       1 |       6 |         540 | {武汉,南昌,长沙}   | (2,1)
       2 |       4 |         579 | {武汉,长沙,南昌}   | (2,2)
       0 |       3 |         832 | {武汉,南昌,长沙,武汉} | (3,0)
       0 |       5 |         832 | {武汉,长沙,南昌,武汉} | (3,0)
       1 |       6 |         866 | {武汉,长沙,南昌,长沙} | (3,1)
(10 rows)
```

如果按照深度优先查找的算法进行排序，使用下面的语句替换递归查询最后的语句。

```
search depth first by city_id set ordercol
select * from trip_journey order by ordercol limit 10;
```

那么结果集按行程逐步扩展，直到找到行程的最大深度。

```
 city_id | trip_id | total_price |   journey_stops    |    ordercol
---------+---------+-------------+--------------------+------------------
       0 |         |           0 | {武汉}             | {(0)}
       1 |       1 |         292 | {武汉,长沙}        | {(0),(1)}
       0 |       3 |         584 | {武汉,长沙,武汉}   | {(0),(1),(0)}
       1 |       1 |         876 | {武汉,长沙,武汉,长沙} | {(0),(1),(0),(1)}
       2 |       2 |         837 | {武汉,长沙,武汉,南昌} | {(0),(1),(0),(2)}
       2 |       4 |         579 | {武汉,长沙,南昌}   | {(0),(1),(2)}
       0 |       5 |         832 | {武汉,长沙,南昌,武汉} | {(0),(1),(2),(0)}
       1 |       6 |         866 | {武汉,长沙,南昌,长沙} | {(0),(1),(2),(1)}
       2 |       2 |         253 | {武汉,南昌}        | {(0),(2)}
       0 |       5 |         506 | {武汉,南昌,武汉}   | {(0),(2),(0)}
(10 rows)
```

前面的结果中会出现多次经过同一个城市的循环情况，PostgreSQL 14 提供了 CYCLE 选项来避免出现该循环。

```
cycle city_id set is_cycle using journey_ids
select * from trip_journey where is_cycle=false;
```

使用 CYCLE 选项，通过检查当前 city_id 是否已经在 journey_ids 列中来标记循环，查询结果通过 is_cycle 过滤。

```
 city_id | trip_id | total_price |  journey_stops   | is_cycle | journey_ids
---------+---------+-------------+------------------+----------+--------------
       0 |         |           0 | {武汉}           | f        | {(0)}
       1 |       1 |         292 | {武汉,长沙}      | f        | {(0),(1)}
       2 |       2 |         253 | {武汉,南昌}      | f        | {(0),(2)}
       2 |       4 |         579 | {武汉,长沙,南昌} | f        | {(0),(1),(2)}
       1 |       6 |         540 | {武汉,南昌,长沙} | f        | {(0),(2),(1)}
(5 rows)
```

SEARCH 选项和 CYCLE 选项提供了一种优雅定义递归查询行为的方法，通过这种方法，用户可以灵活地控制搜索模式并避免 PostgreSQL 14 之前的版本中可能出现的死循环问题。

5.4.6　易用的内置函数引入

PostgreSQL 14 新引入了一些易用的内置函数，string_to_table 函数用于将字符串按分隔符拆分为数据行，与 string_to_array 函数的功能类似，等价于 unnest(string_to_array()) 函数。

string_to_table 函数的示例如下。

```
postgres=# select string_to_table('foo,bar,baz',',');
 string_to_table
-----------------
 foo
 bar
 baz
(3 rows)
```

第一个参数为待拆分的字符串，如果第二个参数的分隔符为 NULL 值，那么字符串的每个字符将作为独立的一行。
```
postgres=# select string_to_table('abcdefg',null);
 string_to_table
-----------------
 a
 b
 c
 d
 e
 f
 g
(7 rows)
```
如果第二个参数的分隔符为 NULL 值，那么整个字符串将作为一行。
```
postgres=# select string_to_table('abcdefg','');
 string_to_table
-----------------
 abcdefg
(1 row)
```
如果第三个参数的分隔符不为 NULL 值，那么匹配的子串将被替换为 NULL 值。
```
postgres=# select string_to_table('ab,cd,ef,gh',',','cd');
 string_to_table
-----------------
 ab

 ef
 gh
(4 rows)
```

注意，string_to_table 函数不适用于解析 CSV 格式字符串。
```
postgres=# select string_to_table('foo,bar,baz,"baz,boo"',',');
 string_to_table
-----------------
 foo
 bar
 baz
 "baz
 boo"
(5 rows)
```

unistr 函数可以将 unicode 转义字符转化为可读的字符串。unicode 转义字符有如下几种表示形式。

- \XXXX（4 位十六进制数）。
- \+XXXXXX（6 位十六进制数）。
- \uXXXX（4 位十六进制数）。
- \UXXXXXXXX（8 位十六进制数）。

下面通过几个示例进行演示。

示例 1 如下。
```
postgres=# select unistr('\0441\043B\043E\043D');
 unistr
```

```
--------
 слон
(1 row)
```

示例 2 如下。

```
postgres=# select
            unistr('d\0061t\+000061'),
            unistr('d\u0061t\U00000061');
 unistr | unistr
--------+--------
 data   | data
(1 row)
```

date_bin 函数可以将指定的时间戳强制截断到最接近指定时间间隔的开头。date_bin 函数的功能与 date_trunc 函数的功能相似，但 date_bin 函数可以将输入的时间戳截断为任意时间间隔，不要求时间间隔只能是一个时间单位。

date_trunc 函数的示例如下。

```
postgres=# select 'untruncated' as spec, now()
        union all
        select spec, date_trunc(spec, now())
        from
 string_to_table('microseconds,milliseconds,second,minute,hour,day,week,month,qua
rter,year,decade,century,millennium',',') as u(spec);
     spec       |             now
----------------+-------------------------------
 untruncated    | 2022-12-08 15:25:30.891473+08
 microseconds   | 2022-12-08 15:25:30.891473+08
 milliseconds   | 2022-12-08 15:25:30.891+08
 second         | 2022-12-08 15:25:30+08
 minute         | 2022-12-08 15:25:00+08
 hour           | 2022-12-08 15:00:00+08
 day            | 2022-12-08 00:00:00+08
 week           | 2022-12-05 00:00:00+08
 month          | 2022-12-01 00:00:00+08
 quarter        | 2022-10-01 00:00:00+08
 year           | 2022-01-01 00:00:00+08
 decade         | 2020-01-01 00:00:00+08
 century        | 2001-01-01 00:00:00+08
 millennium     | 2001-01-01 00:00:00+08
(14 rows)
```

date_bin 函数的示例如下。

```
postgres=# select date_bin('15 minutes', timestamp '2021-05-12 13:41:23', timestamp
   '2001-01-01');
      date_bin
---------------------
 2021-05-12 13:30:00
(1 row)
```

date_bin 函数的示例中的第一个参数为时间间隔，如 15 minutes（minutes 也可简写为 min 或 m）表示以 15min 为间隔，间隔有 4 个时间点，分别为 0min、15min、30min 和

45min。

date_bin 函数的示例中的第二个参数为要处理的时间戳，第三个参数如果有时间部分，那么时间部分将作为偏移量添加到结果的时间部分中。

```
postgres=# select date_bin('15 minutes', timestamp '2021-05-12 13:41:23', timestamp
  '2001-01-01 00:05:01');
     date_bin
---------------------
 2021-05-12 13:35:01
(1 row)
```

按时间间隔截断的 2021-05-12 13:30:00 将添加一个偏移量 05:01，结果是 2021-05-12 13:35:01。

trim_array 函数可以对数组从末尾开始移除指定个数的元素。移除数组末尾的 3 个元素的示例如下。

```
postgres=# select trim_array(array[1, 2, 3, 4, 5, 6], 3);
 trim_array
------------
 {1,2,3}
(1 row)
```

5.5 PostgreSQL 14 的系统层变化

PostgreSQL 14 的系统层变化主要体现在系统元数据引入、系统函数变化、预置角色变化、配置参数变化、客户端 C 驱动改进、客户端认证安全性提高、附加模块变化。

5.5.1 系统元数据引入

PostgreSQL 14 新引入了 5 个系统元数据，如表 5-2 所示。

表 5-2　PostgreSQL 14 新引入的系统元数据

元数据名称	描述
pg_backend_memory_contexts	显示用户会话在服务器不同内存上下文环境下的使用情况，通过 pg_get_backend_memory_contexts 函数实现
pg_stat_progress_copy	跟踪COPY操作执行的进度，包括客户端及服务端的COPY FROM 语句和 COPY TO 语句
pg_stat_replication_slots	显示全局的逻辑复制槽当前状态信息，特别是大事务溢出到磁盘及流式传输的统计信息，通过 pg_stat_get_replication_slot 函数实现。使用 pg_stat_reset_replication_slot 函数可以重置该视图的数据
pg_stat_wal	显示全局 WAL 的活动单行信息，通过 pg_stat_get_wal 函数实现。使用 pg_stat_reset_shared 函数可以重置该视图的数据
pg_stats_ext_exprs	显示扩展统计对象的表达式信息

5.5.2 系统函数变化

PostgreSQL 14 新引入了一些系统函数。

1．新引入了 4 个与开发相关的函数
- string_to_table：将字符串按分隔符拆分为数据行。
- bit_xor：PostgreSQL 14 现已完整支持比特与、比特或、比特异或运算。
- unistr：将 unicode 转义字符转化为可读的字符串。
- trim_array：对数组从末尾开始移除指定个数的元素。

2．新引入了 7 个与范围类型相关的函数
- isempty：检查范围或多范围类型是否为空。
- lower_inc：检查范围或多范围类型的下界是否被包含。
- lower_inf：检查范围或多范围类型的下界是否无穷尽。
- lower：返回范围或多范围类型的下界。
- upper_inc：检查范围或多范围类型的上界是否被包含。
- upper_inf：检查范围或多范围类型的上界是否无穷尽。
- upper：返回范围或多范围类型的上界。

3．新引入了 2 个系统信息函数
- pg_xact_commit_timestamp_origin：同时返回事务的提交时间戳和复制源信息，综合了 pg_xact_commit_timestamp 函数和 pg_last_committed_xact 函数的运行结果。
- pg_get_catalog_foreign_keys：显示系统对象之间的主键与外键的关联信息。

4．新引入了 1 个监控统计函数
- pg_stat_reset_replication_slot：重置 pg_stat_replication_slots 视图复制槽的统计信息，如果函数的输入参数值为 NULL，那么重置所有复制槽的统计信息。

5．新引入了 3 个系统管理函数
- pg_column_compression：获取表的列值 TOAST 支持的压缩算法，如 pglz 算法、lz4 算法。
- pg_log_backend_memory_contexts：将后端进程的内存上下文使用情况记录到数据库日志文件中。
- pg_get_wal_replay_pause_state：显示比 pg_is_wal_replay_paused 函数更加详细的恢复状态。在恢复期间可以查看 3 个值，其中 not paused 代表还未收到暂停请求，pause requested 代表正在执行暂停请求，paused 代表已经执行了暂停请求。

5.5.3 预置角色变化

PostgreSQL 14 新引入了 3 个预置角色，分别是 pg_database_owner、pg_read_all_data 和 pg_write_all_data。

pg_database_owner 表示当前数据库的宿主，常用于在模板数据库中进行一些通用的定制化创建工作，并将权限赋予当前数据库的宿主而不是某个用户，之后新建的数据库不再需要手动赋权。

例如，要提高 public 模式的安全性，不允许普通用户在 public 模式下创建对象，只有数据库的宿主才能创建对象，可以先在 template1 数据库中进行如下设置。

```
revoke all on schema public from postgres;
revoke all on schema public from public;
alter schema public owner to pg_database_owner;
grant all on schema public to pg_database_owner;
grant usage on schema public to public;
```

在上述代码中，前两行用于回收 postgres 角色与 public 角色的权限，第三行用于修改 public 模式的宿主，最后两行用于重新设置 public 模式的权限，数据库的宿主对 public 模式有全部权限，public 角色则取消了创建权限，只有使用权限。

下面将新建的 mydb1 数据库的宿主设置为 user1，其他用户在 public 模式下默认没有创建的权限，此时创建表会报错。

```
template1=# create database mydb1 owner user1;
CREATE DATABASE

template1=# \c mydb1 user2
You are now connected to database "mydb1" as user "user2".

mydb1=> create table public.t(id int);
ERROR: permission denied for schema public
LINE 1: create table public.t(id int);
```

pg_read_all_data 可以用来创建具有只读权限的用户，提供实例级全局读取权限。在 PostgreSQL 14 之前的版本中只能对单个数据库分别设置当前所有表的只读权限和新建表的默认权限，并且不同的数据库对象需要分类设置。被赋予 pg_read_all_data 的用户将具有所有表、视图及序列的查询权限。

```
grant pg_read_all_data to user_read;
```

pg_write_all_data 可以用来创建具有写入权限的用户，提供实例级全局写入权限。被赋予 pg_write_all_data 的用户将具有插入、更新和删除所有表、视图及序列的权限。

```
grant pg_write_all_data to user_write;
```

PostgreSQL 14 支持的预置角色如表 5-3 所示。

表 5-3　PostgreSQL 14 支持的预置角色

预置角色名称	描述
pg_database_owner	提供数据库宿主的通用权限
pg_read_all_data	提供全局读取的访问权限
pg_write_all_data	提供全局写入的访问权限
pg_execute_server_program	可以执行数据库服务端程序以配合 COPY 操作和其他允许执行服务端程序的函数
pg_read_server_files	可以使用 COPY 操作及其他文件访问函数在数据库服务端可访问的任意位置读取文件

续表

预置角色名称	描述
pg_write_server_files	可以使用 COPY 操作及其他文件访问函数在数据库服务端可访问的任意位置写入文件
pg_read_all_settings	可以读取所有配置变量，包括那些通常只对超级用户可见的信息
pg_read_all_stats	可以读取所有 pg_stat_*视图信息，以及相关的扩展统计信息，包括那些通常只对超级用户可见的信息
pg_stat_scan_tables	可以执行一些监控函数，需要获取表的 Access Share 锁
pg_monitor	可以执行各种监视视图和函数，是 pg_read_all_settings、pg_read_all_stats 和 pg_stat_scan_tables 的成员
pg_signal_backend	可以使用 pg_cancel_backend 函数或 pg_terminate_backend 函数对后端进程发送信号，取消后端进程查询或中止后端进程

5.5.4 配置参数变化

PostgreSQL 14 中配置参数的主要变化有：新引入了 17 个配置参数，修改了 3 个配置参数，并废弃了 2 个配置参数，如表 5-4～表 5-6 所示。

表 5-4 PostgreSQL 14 新引入的配置参数

参数名称	描述
client_connection_check_interval	检测客户端是否连接，如果客户端已离线，那么可以快速中断运行中的长 SQL 语句
compute_query_id	设置由系统生成查询语句的唯一标识
debug_discard_caches	参数 debug，用于加快系统缓存刷写速度
default_toast_compression	设置全局 TOAST 支持的压缩算法，如 pglz 算法、lz4 算法
enable_async_append	控制规划器使用 async append
enable_memoize	控制规划器使用 memoization，用于 nested loop join 对结果进行缓存
huge_page_size	在数据库层设置大页的尺寸，默认值为 0，在系统层设置
idle_session_timeout	与参数 idle_in_transaction_session_timeout 的功能类似，但使用本参数杀死的不是正在事务中的会话，而是空闲的会话
in_hot_standby	报告数据库是否处于 hot standby 状态。本参数是一个只读参数，可以发送到客户端，以使其更容易检测状态
log_recovery_conflict_waits	控制 standby 模式恢复冲突是否记录等待超时
min_dynamic_shared_memory	允许在数据库启动时预设动态内存
recovery_init_sync_method	设置数据库崩溃恢复前同步数据文件调用的系统层同步方法
remove_temp_files_after_crash	控制数据库后端用户进程崩溃后删除临时文件。之前保留临时文件是为了方便调试，如果需要保留那么可以关闭本参数

参数名称	描述
ssl_crl_dir	设置 SSL 吊销证书的目录
track_wal_io_timing	采集 WAL I/O 活动的时间花费
vacuum_failsafe_age	采用紧急 VACCUM 模式来更好地保护事务 ID 回卷
vacuum_multixact_failsafe_age	与参数 vacuum_failsafe_age 的功能类似,用于子事务环境中

表 5-5 PostgreSQL 14 修改的配置参数

参数名称	描述
checkpoint_completion_target	默认值由 0.5 变为 0.9,让检查点执行得更加平滑
password_encryption	默认值由 MD5 变为 SCRAM-SHA-256,提高客户端认证的安全性
vacuum_cost_page_miss	默认值由 10 变为 2,根据现阶段硬件的环境进行降低

表 5-6 PostgreSQL 14 废弃的配置参数

参数名称	描述
operator_precedence_warning	用于处理迁移问题的兼容性参数,随着大版本策略的出现而被废弃
vacuum_cleanup_index_scale_factor	因对 AUTOVACUUM 进程有不利影响而被废弃

5.5.5 客户端 C 驱动改进

基于 ODBC 接口或 Python 语言编程,将使用 C 驱动的 libpq 协议来连接数据库,在 PostgreSQL 14 中对 C 驱动的 libpq 协议做了一些改进,主要包括以下两个方面。

- 新增管道模式:管道模式允许应用程序发送查询语句而无须读取先前发送的查询结果。利用管道模式,客户端等待服务端的时间会更少,当服务端距离较远且网络延时较高时使用管道模式比较合适。
- 完善参数 target_session_attrs 的值:新增 read-only、primary、standby、prefer-standby 几个值。

客户端有两种常见的协议,分别是 jdbc 协议和 libpq 协议。jdbc 协议可以使用参数 targetServerType 连接特定状态的数据库节点,可选值包括 any、primary、master、slave、secondary、preferSlave 和 preferSecondary。libpq 协议可以使用参数 target_session_attrs 连接特定状态的数据库节点,在 PostgreSQL 14 之前的版本中,参数 target_session_attrs 只支持如下两个值。

- any(默认值):允许连接到任意数据库节点,从所有配置的连接中轮流尝试,直至连接建立成功,从而实现故障转移。
- read-write:在连接时,只接收可以读写的数据库节点。在建立连接后,会发送 SHOW transaction_read_only,若为 on,则代表是只读库,会把连接关闭,并测试第二个数据库,以此类推,直至连接到支持读写的数据库节点。

PostgreSQL 14 中的参数 target_session_attrs 参照 jdbc 协议的参数 targetServerType 的值，增加了 4 个值。完整的参数 target_session_attrs 的值如下。

- any（默认值）：允许连接到任意数据库节点。
- read-write：仅接收连接的 default_transaction_read_only=off 并且非 standby 节点。
- read-only：仅接收连接的 default_transaction_read_only=on 或 standby 节点。
- primary：仅接收连接的非 standby 节点。
- standby：仅接收连接的 standby 节点。
- prefer-standby：尝试连接 standby 节点，如果连接失败那么尝试连接任意数据库节点。

从 PostgreSQL 14 开始，参数 target_session_attrs 的值与参数 targetServerType 的值一一对应。

下面对参数 target_session_attrs 进行测试，在本机环境下对以下两个值进行测试。

首先测试 read-only，仅连接只读事务环境的节点。

```
$ psql "target_session_attrs=read-only host=localhost,localhost port=5432,6432 dbname=postgres"
psql: error: connection to server at "localhost" (127.0.0.1), port 5432 failed: session is not read-only
connection to server at "localhost" (127.0.0.1), port 6432 failed: session is not read-only
```

可以看到，连接失败，这是因为两个节点的参数 default_transaction_read_only 的值默认是 off。

其次，测试 standby，仅连接 standby 节点。

```
$ psql "target_session_attrs=standby host=localhost,localhost port=5432,6432 dbname=postgres"
psql: error: connection to server at "localhost" (127.0.0.1), port 5432 failed: server is not in hot standby mode
connection to server at "localhost" (127.0.0.1), port 6432 failed: server is not in hot standby mode
```

可以看到，连接的两个节点都未处于 standby 模式，连接失败。

下面修改 6432 节点，添加 standby 触发文件。

```
$ touch /opt/pgdata6432/standby.signal
```

最后，启动服务器进行测试。

```
$ psql "target_session_attrs=standby host=localhost,localhost port=5432,6432 dbname=postgres"
psql (14.6)
Type "help" for help.

postgres=#
```

可以看到，连接成功，参数 target_session_attrs 不仅可以按照期望的节点进行连接，而且可以很方便地动态扩展多个节点。

5.5.6 客户端认证安全性提高

MD5 是 PostgreSQL 10 之前的版本中唯一可用的口令加密选项，其存储方式为

MD5('密码+用户名'),容易出现字典攻击和泄露问题。PostgreSQL 10 开始支持 SCRAM-SHA-256 方式,SCRAM-SHA-256 方式的使用使得破解密码变得非常困难。

很多用户在升级到 PostgreSQL 14 之后,发现应用程序接口提示数据库连接失败,这是由 PostgreSQL 14 默认强制使用 SCRAM-SHA-256 方式,而客户端使用 PostgreSQL 14 之前的版本的驱动不支持 SCRAM-SHA-256 方式引起的。

```
postgres=# show password_encryption;
 password_encryption
---------------------
 scram-sha-256
(1 row)
```

使用 PostgreSQL 10 到 PostgreSQL 14,可以修改参数 password_encryption 的值来提升安全性。同时,注意要及时升级到支持 SCRAM-SHA-256 方式的客户端驱动版本。

5.5.7 附加模块变化

在 PostgreSQL 14 中附加模块有两个新增的扩展插件。

第一个新增的扩展插件是 old_snapshot。old_snapshot 插件用于通过 pg_old_snapshot_time_mapping 函数探测每分钟 XID 的映射状态。

当关闭参数 old_snapshot_threshold(默认值为-1,表示不限制事务快照),限制事务快照时,可以通过 SQL 接口进行观测。

```
postgres=# show old_snapshot_threshold;
 old_snapshot_threshold
------------------------
 10min
(1 row)

postgres=# select end_timestamp,clock_timestamp(),newest_xmin,age(newest_xmin)
   from pg_old_snapshot_time_mapping();
     end_timestamp      |        clock_timestamp         | newest_xmin | age
------------------------+--------------------------------+-------------+------
 2022-11-19 17:35:00+08 | 2022-11-19 17:53:02.181933+08  |      918515 | 1416
 2022-11-19 17:36:00+08 | 2022-11-19 17:53:02.181957+08  |      918516 | 1415
 2022-11-19 17:37:00+08 | 2022-11-19 17:53:02.181959+08  |      918516 | 1415
 2022-11-19 17:38:00+08 | 2022-11-19 17:53:02.181961+08  |      918516 | 1415
 2022-11-19 17:39:00+08 | 2022-11-19 17:53:02.181964+08  |      918527 | 1404
 2022-11-19 17:40:00+08 | 2022-11-19 17:53:02.181966+08  |      918828 | 1103
 2022-11-19 17:41:00+08 | 2022-11-19 17:53:02.181968+08  |      919132 |  799
 2022-11-19 17:42:00+08 | 2022-11-19 17:53:02.18197+08   |      919135 |  796
 2022-11-19 17:43:00+08 | 2022-11-19 17:53:02.181973+08  |      919135 |  796
 2022-11-19 17:44:00+08 | 2022-11-19 17:53:02.181975+08  |      919436 |  495
 2022-11-19 17:45:00+08 | 2022-11-19 17:53:02.181977+08  |      919439 |  492
 2022-11-19 17:46:00+08 | 2022-11-19 17:53:02.181979+08  |      919439 |  492
 2022-11-19 17:47:00+08 | 2022-11-19 17:53:02.181981+08  |      919439 |  492
 2022-11-19 17:48:00+08 | 2022-11-19 17:53:02.181983+08  |      919439 |  492
 2022-11-19 17:49:00+08 | 2022-11-19 17:53:02.181986+08  |      919439 |  492
 2022-11-19 17:50:00+08 | 2022-11-19 17:53:02.181988+08  |      919439 |  492
```

```
 2022-11-19 17:51:00+08 | 2022-11-19 17:53:02.18199+08  |          919439 |  492
 2022-11-19 17:52:00+08 | 2022-11-19 17:53:02.181992+08 |          919439 |  492
 2022-11-19 17:53:00+08 | 2022-11-19 17:53:02.181994+08 |          919668 |  263
 2022-11-19 17:54:00+08 | 2022-11-19 17:53:02.181996+08 |          919911 |   20
(20 rows)
```

第二个新增的扩展插件是 pg_surgery。pg_surgery 插件用于纠正损坏的数据且更改行可见性，使删除的元组重新出现。pg_surgery 插件提供了以下两个函数。

- heap_force_freeze：强制冻结元组。
- heap_force_kill：强制删除元组。

其中，heap_force_freeze 函数与 pg_dirtyread 插件可以搭配使用。

下面通过示例进行演示。

创建扩展插件及测试表。

```
create extension pg_dirtyread ;

create extension pg_surgery ;

create table test_undelete (id int, t text);

alter table test_undelete set (
  autovacuum_enabled = false, toast.autovacuum_enabled = false
);
```

使用 pg_dirtyread 插件能从表中读取死元组，即已经删除或更新的行，但是该插件只在表没有被清理时才有效。因此，上面代码中关闭了 test_undelete 表的自动清理参数。

插入 3 条数据。

```
postgres=# insert into test_undelete values (1,'data1'),(2,'data2'),(3,'data3');
INSERT 0 3

postgres=# select * from test_undelete ;
 id |   t
----+-------
  1 | data1
  2 | data2
  3 | data3
(3 rows)
```

删除 id 属性的值小于 3 的两条记录。

```
postgres=# delete from test_undelete where id < 3;
DELETE 2

postgres=# select * from test_undelete ;
 id |   t
----+-------
  3 | data3
(1 row)
```

此时，前两条数据已经消失，下面使用 pg_dirtyread 插件结合 heap_force_freeze 函数进行恢复。

```
postgres=# select heap_force_freeze('test_undelete', array_agg(ctid))
            from pg_dirtyread('test_undelete')
t(ctid tid, xmin xid, xmax xid, id int, t text)
where id = 1;
 heap_force_freeze
-------------------

(1 row)

postgres=# select * from test_undelete;
 id |   t
----+-------
  1 | data1
  3 | data3
(2 rows)
```

heap_force_freeze 函数通过 ctid 列强制冻结元组和标记为未删除来恢复数据。从上面的代码中可以看出，第一条数据已经恢复。

使用 pg_dirtyread 插件可以查看完整的 3 条数据，并显示行位置和可见性等系统列。

```
postgres=# select * from pg_dirtyread('test_undelete') t(ctid tid, xmin xid, xmax
   xid, id int, t text);
 ctid  |  xmin  |  xmax  | id |   t
-------+--------+--------+----+-------
 (0,1) |      2 |      0 |  1 | data1
 (0,2) | 918512 | 918513 |  2 | data2
 (0,3) | 918512 |      0 |  3 | data3
(3 rows)
```

如果表已被系统清理，那么使用上述方法无效。注意，由于 pg_dirtyread 插件和 pg_surgery 插件运行于正常的 MVCC 机制之外，因此它们很容易创建损坏的数据，包括重复的行、重复的主键值、不同步的索引、被损坏的外键约束等。如果 pg_dirtyread 插件和 pg_surgery 插件使用不当，那么很容易损坏以前未损坏的数据，加快损坏数据库的速度。因此，必须谨慎使用 pg_dirtyread 插件和 pg_surgery 插件。

5.6 本章小结

PostgreSQL 14 在性能提升方面主要包括在大量连接场景下获取事务快照瓶颈优化，活动连接和空闲连接的扩展性得到明显的改善；引入紧急清理模式，可以更好地预防事务 ID 回卷；列级压缩可配置，新增 lz4 算法，可以有效地提高压缩和解压缩的速度，lz4 算法的压缩性能比 pglz 算法好；逻辑复制的大量性能得到改进，包括大事务的流式处理、TRUNCATE 操作的性能提升、使用二进制形式传输数据、表初始数据同步优化等。

PostgreSQL 14 在可靠性提高方面主要包括 amcheck 模块支持堆表数据逻辑错误检测功能；备节点可作为恢复源，以减轻主库压力；密码长度限制取消。

PostgreSQL 14 在运维管理优化方面主要包括引入查询 ID，方便分析慢查询；重建索

| 快速掌握 PostgreSQL 版本新特性 |

引,支持在线移动索引表空间;支持在线重建触发器;引入后台操作进度报告;增加可观测性视图。PostgreSQL 14 在开发易用性提升方面主要包括引入多范围类型;支持在存储过程中直接使用 OUT 模式参数;引入新形式的 SQL 函数,可以检测依赖性;JSON 操作功能增强,JSON 数据类型支持数组下标的访问形式等。PostgreSQL 14 在系统层变化方面主要包括客户端 C 驱动改进、客户端认证安全性提高等。

第 6 章

PostgreSQL 15 新特性

6.1　PostgreSQL 15 的主要性能提升

与 PostgreSQL 15 之前的版本相比，PostgreSQL 15 的性能有了一定的提升，主要体现在统计信息内存化、增量排序算法改进、WAL 恢复预读取、全页写新增压缩算法、备份效率提高、并行特性增强。

6.1.1　统计信息内存化

在 PostgreSQL 中有一个统计信息系统，该系统通过 stats collector 进程采集本地服务器上的一些活动信息，如表或索引的访问次数，或最近一次清理、自动清理的时间及运行的次数等。统计信息系统收集的数据通过不同的 pg_stat_*视图展示给用户。

在 PostgreSQL 15 中，会发现 stats collector 进程消失了。

```
8685 ?        Ss     0:00 /opt/pg15/bin/postgres -D /opt/pgdata1500
8686 ?        Ss     0:00  \_ postgres: checkpointer
8687 ?        Ss     0:00  \_ postgres: background writer
8689 ?        Ss     0:00  \_ postgres: walwriter
8690 ?        Ss     0:00  \_ postgres: autovacuum launcher
8691 ?        Ss     0:00  \_ postgres: logical replication launcher
```

在 PostgreSQL 14 及 PostgreSQL 14 之前的版本中，stats collector 进程是伴随着主进程启动的。

```
3819 ?        Ss     0:01 /opt/pg14/bin/postgres -D /opt/pgdata1405
```

```
3820 ?        Ss    0:00  \_ postgres: logger
3825 ?        Ss    0:00  \_ postgres: checkpointer
3826 ?        Ss    0:00  \_ postgres: background writer
3827 ?        Ss    0:01  \_ postgres: walwriter
3828 ?        Ss    0:01  \_ postgres: autovacuum launcher
3830 ?        Ss    0:01  \_ postgres: stats collector
3833 ?        Ss    0:00  \_ postgres: logical replication launcher
```

虽然在 PostgreSQL 14 及 PostgreSQL 14 之前的版本中确实会存在因 hosts 文件配置错误导致 stats collector 进程和 autovacuum launcher 进程启动失败的问题，但这里并非因为 PostgreSQL 15 或系统层存在错误配置。

PostgreSQL 中的每个后端会话都是一个独立的进程，统计信息通过不同的进程进行传输通信并不是一件容易的事情。每个后端会话都采用 UDP 方式将各自的活动信息统一发送到 stats collector 进程中，这种方式存在一些问题。首先，受限于单一的 stats collector 进程，经常可以看到一些与之相关的故障，如因收集较慢导致信息陈旧，或因 hosts 文件配置错误导致不能正常启动进程等。此外，如果打开调试级别的日志，那么可能会从日志中看到频繁地写信息，这会导致数据目录所在的挂载点产生大量的 I/O。这是因为 stats collector 进程存储数据使用的默认目录位于数据目录中。

```
postgres=# show stats_temp_directory;
 stats_temp_directory
----------------------
 pg_stat_tmp
(1 row)
```

从 PostgreSQL 15 开始，统计信息不再通过独立的系统收集，而以纯内存的实现方式直接集成到内核系统中，使用共享内存而不是临时文件存储统计信息，并且即使在重启服务后也能加载回来。这是因为统计信息在数据库关闭前由检查点进程将其持久化写入文件系统，并在启动服务时由启动进程将其加载。这种实现方式更加可靠，无须通过 UDP 方式进行通信，减少了 I/O 和进程之间的通信，从而改进了性能。

随着统计信息系统所有开销及维护工作的消失，其他系统如 AUTOVACUUM 进程要做的工作就更少了。此外，用于经常查询统计信息的监控工具也会大大降低系统负载。

注意，参数 stats_temp_directory 虽然因为统计信息内存化而被弃用，但是保留目录 pg_stat_tmp 或环境变量 PG_STAT_TMP_DIR 可以兼容 pg_stat_statements 插件的依赖。

6.1.2 增量排序算法改进

为了让排序速度更快，近几年来，各个版本的 PostgreSQL 都在对排序进行优化。当在 PostgreSQL 中运行应用程序时，除了可以直接使用 ORDER BY 子句进行排序，还可以通过以下几种方式进行排序。
- 使用 ORDER BY 子句的聚合函数。
- 使用 GROUP BY 子句。
- 使用产生 Merge Join 执行计划的查询。

- 使用 UNION 查询。
- 使用 DISTINCT 查询。
- 使用带有 PARTITION BY 子句及 ORDER BY 子句的窗口函数查询。

下面介绍 PostgreSQL 15 中的几项排序优化策略。

- 优化一：单列排序性能优化。

在 PostgreSQL 15 之前的版本中，执行器在排序期间会存储整个元组，而在 PostgreSQL 15 中查询单列排序场景时仅存储单列数据，仅存储单列数据也就意味着无须将整个元组复制到内存中。这个特性比较容易出现在 EXISTS 子句或 NOT EXISTS 子句的查询中。

- 优化二：使用"generation"分配器降低内存消耗。

在分配排序所需的内存时，PostgreSQL 15 之前的版本使用"aset"分配器，"aset"分配器请求的内存大小会向上取整为 2 的下一次幂。例如，24 字节的分配请求会变为 32 字节，而 600 字节的分配请求会变为 1024 字节。"aset"分配器的这种按二次幂取整的规则会导致平均 25%的内存浪费。PostgreSQL 15 可以选用新的"generation"分配器，它采用先进先出的分配模式，不会向上取整分配，能以更少的内存存储更多的信息。

- 优化三：为常见的数据类型添加专门的排序例程。

PostgreSQL 中有很多的数据类型，每种数据类型都有相应的比较函数，用于排序时两个值的比较。虽然这些比较函数采用 C 语言编写，调用开销很低，但是多次调用会产生明显的开销。PostgreSQL 15 添加了一组新的快速排序函数来适配一些常见的数据类型，这些快速排序函数内部实现通过内联属性的比较函数来降低调用开销。常见的 integer、timestamp 及使用默认的 C 语言排序规则的文本等数据类型均被惠及。

- 优化四：Merge 排序算法使用多路归并（K-way Merge）代替多阶段合并（Polyphase Merge）。

针对超过参数 work_mem 的值的大型排序，Merge 排序算法使用多路归并代替多阶段合并，以降低 I/O 的消耗。

6.1.3　WAL 恢复预读取

PostgreSQL 使用 WAL 记录数据变化，表或索引的持久化对象文件的修改落盘是后台异步进行的，用户在前台提交事务后只需要等待 WAL 落盘即可。这种统筹设计能提高性能，用户只需要等待 WAL 顺序写入，在异常情况下的恢复也是可靠的。

WAL 的恢复有以下两种常见场景。

- 崩溃重启，恢复最近一次检查点以来的所有变化。
- 物理流复制场景下的持续重放恢复。

WAL 的记录结构比较简单，依序重放 WAL 的恢复过程虽然对 CPU 的影响不大，但会带来 I/O 延迟问题，如当从磁盘中获取需要修改的数据时恢复进程会休眠，在某个时刻数据块落盘时也需要等待。

PostgreSQL 15 对 WAL 恢复期间的 I/O 延迟问题进行了优化，新增了参数 recovery_prefetch 进行预读取。这个预读取功能指通过在恢复期间尝试预读取 WAL 中引用且尚未缓存的数据块来缩短 I/O 的等待耗时。

为了更好地体验这个特性，可以进行如下设置。
- 调整参数 maintenance_io_concurrency 的值。
- 监测 pg_stat_recovery_prefetch 视图。
- 尝试增加参数 wal_decode_buffer_size 的值。

6.1.4 全页写新增压缩算法

全页写（Full Page Write）发生于检查点之后的第一次修改中，PostgreSQL 支持在设置参数 wal_compression 的值为 on 后采用内置的 pglz 算法压缩全页写再写入 WAL 的特性。

该特性默认关闭，早在 2019 年就有人在 PostgreSQL 9.5 中修改参数 wal_compression 的值为 on，对比测试发现，可以节省 50%的空间和提高带宽，但 CPU 会变得繁忙一些，这是因为需要进行加密压缩。

在打开参数 wal_log_hints 来使用 pg_rewind 工具时，WAL 的总量会显著增加，此时优化参数 wal_compression 会有明显的收获。另外，如果 WAL 产生的速度较快，同时还有归档文件，那么在运行速度比较慢的磁盘上会面临追赶不上 WAL 生成速度的情况，更复杂的情况是选择了较低带宽的网络。在这些情况下，打开参数 wal_compression 虽然会增加 CPU 的负载（包括记录 WAL 的压缩过程和应用回放 WAL 的解压过程），但会减少 I/O 负载，同时会减轻网络压力，对现代硬件而言，CPU 的压力不再是什么大问题。

在 PostgreSQL 15 之前的版本中，参数 wal_compression 的值只能被设置为 off 和 on，PostgreSQL 15 除了可以设置 off 和 on 两个值，还新增了 pglz、lz4 和 zstd 三个值。基准测试表明，lz4 算法及 zstd 算法的压缩效率能轻松超越 pglz 算法的压缩效率，zstd 算法在某些场景下可以调低默认的压缩级别从而减少 CPU 的消耗。对于通用场景，还是推荐使用默认的压缩级别。

6.1.5 备份效率提高

pg_basebackup 工具是一个使用非常简便的基础备份工具，使用它可以对整个实例进行物理备份。pg_basebackup 工具常用于搭建流复制环境或恢复新的备份实例。与 PostgreSQL 15 之前的版本相比，PostgreSQL 15 的基础备份效率有了一定的提高。

- 新增服务端压缩支持：以前只能控制 pg_basebackup 工具在客户端进行压缩，现在可以控制 pg_basebackup 工具在客户端或在服务端进行压缩。
- 新增 lz4 算法和 zstd 算法：以前只能使用 gzip 算法，现在服务端及客户端都支持 gzip 算法、lz4 算法和 zstd 算法。要使用 lz4 算法和 zstd 算法需要在编译时打开 --with-lz4 选项和--with-zstd 选项。
- 使用--compress 选项可以指定压缩方法及压缩级别。
- 在使用 zstd 算法压缩时，可以使用多线程 workers 并行处理，PostgreSQL 15 的 zstd 库支持并行压缩，既可以在客户端进行压缩，又可以在服务端进行压缩。

- 使用文本备份方式可以在服务端压缩后传输到客户端解压缩。

使用 pg_basebackup 工具对--compress 选项指定压缩方法和压缩级别。

```
--compress=[{client|server}-]METHOD[:DETAIL]
```

PostgreSQL 15 之前的版本只能设置压缩级别,压缩级别通过在冒号后直接使用数字或 level=number 来指定。对于 zstd 算法压缩,还可以设置并行处理的工作线程数。

--compress 选项的具体使用示例如下。

- 使用文本格式,不压缩。

```
$ pg_basebackup -Fp --compress=none -D data
```

- 使用 tar 包格式,gzip 算法压缩。

```
$ pg_basebackup -Ft --compress=gzip -D data

$ pg_basebackup -Ft --compress=client-gzip -D data
```

上面两种方式类似,进行客户端 gzip 算法压缩时可以使用 client-gzip,或简化为 gzip。

```
$ pg_basebackup -Ft --compress=server-gzip -D data

$ pg_basebackup -Ft --compress=server-gzip:5 -D data

$ pg_basebackup -Ft --compress=server-gzip:level=5 -D data
```

在进行服务端 gzip 算法压缩时可以使用 server-gzip,压缩级别可以直接在冒号后添加数字,或使用 level=5 来指定,压缩级别的范围是 1~9,数字越大,压缩率越高。

- 使用 tar 包格式,lz4 算法压缩。

使用 lz4 算法进行压缩比使用 gzip 算法进行压缩的速度要快一些,压缩率要低一些,如果期望速度更快,那么可以使用 lz4 算法代替 gzip 算法。

```
$ pg_basebackup -Ft --compress=client-lz4 -D data

$ pg_basebackup -Ft --compress=server-lz4:12 -D data

$ pg_basebackup -Ft --compress=server-lz4:level=12 -D data
```

在进行服务端 lz4 算法压缩时可以使用 server-lz4,压缩级别的写法与前面类似,也是两种写法,压缩级别的范围是 1~12,数字越大,压缩率越高。

- 使用 tar 包格式,zstd 算法压缩。

zstd 算法压缩结合了 gzip 算法与 lz4 算法压缩的优点,不仅压缩速度快,而且压缩率高,同时还支持使用多个 zstd workers 进程并行备份数据,如果 CPU 的内存比较充足,那么 zstd 算法压缩是很好的性能提升方式。

```
$ pg_basebackup -Ft --compress=server-zstd:level=5,workers=2 -D data
```

- 使用文本格式,lz4 算法压缩。

有些用户不喜欢使用 tar 包格式,这是因为 tar 包格式无法直接使用 pg_verifybackup 工具来校验备份数据的完整性,此时使用文本格式也能在服务端进行压缩,在客户端自动进行解压缩。

```
$ pg_basebackup -Fp --compress=server-lz4 -D data
```

在使用文本格式时,虽然压缩并没有给用户带来任何性能收益,但是可以高效地传输数据,加速服务端的备份数据从网络上传输到远程客户端。当网络带宽较低时,使用这种方式比较合适。

总体来说，在服务端进行压缩比在客户端进行压缩的速度要慢一些，在网络带宽较低的情况下，可以先在服务端进行压缩再传输到客户端。用户可以根据是想优先减少备份压缩的时间还是优化备份存储占用的空间对 PostgreSQL 15 提供的多种方式进行选择。

6.1.6 并行特性增强

自 PostgreSQL 9.6 引入并行查询特性以来，各版本陆续完善并行特性。
- PostgreSQL 9.6 支持并行全表扫描、并行 Aggregate、并行 JOIN。
- PostgreSQL 10 支持并行 B-Tree 索引扫描、Bitmap 扫描及 Merge Join，并支持并行 PREPARE 语句和 EXECUTE 语句。
- PostgreSQL 11 支持 B-Tree 索引并行创建、HASH 操作及 HASH JOIN 操作并行。
- PostgreSQL 13 支持单表的索引并行清理。
- PostgreSQL 14 支持 RETURN QUERY 语句及 postgres_fdw 模块使用并行。

PostgreSQL 15 新增支持 DISTINCT 操作并行，下面通过示例进行演示。

创建测试表并插入数据。

```
create table t1 ( a int );
insert into t1 select 1 from generate_series(1,100000);
insert into t1 select 2 from generate_series(1,300000);
insert into t1 select 4 from generate_series(1,400000);
insert into t1 select 5 from generate_series(1,500000);
analyze t1;
```

在 PostgreSQL 14 中，查询 a 字段进行 DISTINCT 操作的实际执行计划。

```
postgres=# explain analyze select distinct a from t1;
                                QUERY PLAN
------------------------------------------------------------------------
 HashAggregate  (cost=22003.00..22003.04 rows=4 width=4) (actual
   time=808.701..808.703 rows=4 loops=1)
   Group Key: a
   Batches: 1  Memory Usage: 24kB
   -> Seq Scan on t1  (cost=0.00..18753.00 rows=1300000 width=4) (actual
     time=0.008..240.300 rows=1300000 loops=1)
 Planning Time: 0.117 ms
 Execution Time: 808.950 ms
(6 rows)
```

在 PostgreSQL 15 中，执行同样的操作。

```
postgres=# explain analyze select distinct a from t1;
                                QUERY PLAN
------------------------------------------------------------------------
 Unique  (cost=13524.79..13524.83 rows=4 width=4) (actual time=382.621..386.055
   rows=4 loops=1)
   -> Sort  (cost=13524.79..13524.81 rows=8 width=4) (actual time=382.538..385.965
     rows=12 loops=1)
       Sort Key: a
       Sort Method: quicksort  Memory: 25kB
```

```
               -> Gather  (cost=13523.83..13524.67 rows=8 width=4)
(actual time=382.402..385.937 rows=12 loops=1)
              Workers Planned: 2
              Workers Launched: 2
                -> HashAggregate  (cost=12523.83..12523.87 rows=4 width=4) (actual
   time=345.671..345.674 rows=4 loops=3)
                    Group Key: a
                    Batches: 1  Memory Usage: 24kB
                    Worker 0:  Batches: 1  Memory Usage: 24kB
                    Worker 1:  Batches: 1  Memory Usage: 24kB
                    -> Parallel Seq Scan on t1  (cost=0.00..11169.67 rows=541667
   width=4) (actual time=0.029..115.530 rows=433333 loo
 Planning Time: 0.292 ms
 Execution Time: 386.237 ms
(15 rows)
```

从执行计划中可以看出，DISTINCT 操作可以分为两个阶段，第一个阶段通过多个 workers 进程发起并行扫描，第二阶段通过 SORT 和 UNIQUE 去除组合之后的结果。

6.2　PostgreSQL 15 的可靠性提高

PostgreSQL 15 的可靠性提高主要体现在统计信息一致性读取、统一非独占备份模式引入、本地化 Collation 相关增强、流复制支持 UNLOGGED 序列、pg_rewind 工具指定外部配置文件。

6.2.1　统计信息一致性读取

PostgreSQL 15 中的统计信息采用新的共享内存架构，当统计信息并发更新时，读取的一致性就成了问题。PostgreSQL 15 新引入了一个参数 stats_fetch_consistency，该参数用于控制一个事务中多次访问统计信息的行为。

使用参数 stats_fetch_consistency 控制一致性读取的值包括 none、cache 和 snapshot。
- none：每次从共享内存中重新获取，效率最高，适合监控类软件使用的查询。
- cache：首次访问后缓存，直到事务结束或手动调用 pg_stat_clear_snapshot 函数才会重新获取。cache 可以确保重复获取相同的值，效率适中，适合自连接等查询。
- snapshot：首次访问后 Database 级别缓存，直到 transaction 结束或手动调用 pg_stat_clear_snapshot 函数才会重新获取。snapshot 可以确保在当前数据库中获取相同的值，费用最高。

参数 stats_fetch_consistency 的默认值为 cache，用户可以根据不同的使用场景进行调整。

6.2.2 统一非独占备份模式引入

在介绍备份模式之前，先介绍一下备份函数在 PostgreSQL 15 中的变化。在使用自制备份脚本进行基础备份时，PostgreSQL 15 之前的版本使用下面的两个函数。
- pg_start_backup 函数。
- pg_stop_backup 函数。

从 PostgreSQL 15 开始，上面两个函数被下面两个函数代替。
- pg_backup_start 函数。
- pg_backup_stop 函数。

这样更换之后，不仅方便进行 Tab 键补全，而且容易搜索和识记，特别是可以区分独占备份模式。仔细观察 PostgreSQL 15 中的 pg_backup_start 函数和 pg_backup_stop 函数的参数列表可以发现，参数 exclusive 消失了。

一直以来，独占备份模式并不被推荐使用。使用它带来的最大问题是如果服务器在这种模式下突然停止运行，那么之后服务器可能无法正常启动。这是因为在独占备份模式下，系统会在数据目录中创建一个备份标签文件，其中包含启动备份的最新检查点位置。这样能确保在启动服务器时不会从控制文件中读取检查点位置。

使用独占备份模式可能会导致数据损坏，这是因为备份可能包含该检查点之前的数据文件。如果系统业务比较繁忙，那么 pg_wal 目录中的 WAL 文件可能已经被归档和回收，随后服务器将启动失败，此时只有手动删除备份标签文件才能重启服务器。此外，如果使用参数 restore_command 来获取所需的 WAL，那么很容易混淆实例到底是从崩溃恢复的还是从备份恢复的。

PostgreSQL 15 删除了一直被弃用的独占备份模式，在非独占备份模式下通过 pg_backup_stop 函数返回检查点及表空间映射等信息，而不在数据目录中创建。注意，在备份期间需要保持数据库连接的连续性，如果客户端在进行备份时断开连接那么备份将被中止。此外，与独占备份模式相关的 pg_is_in_backup 函数和 pg_backup_start_time 函数也将被删除。

6.2.3 本地化 Collation 相关增强

在 PostgreSQL 中，LOCALE 默认使用 C 的本地化排序规则。在操作系统和各种软件中也经常可以看到 LOCALE 的相关配置。LOCALE 是一种文化偏好的区域设置，包括字母表、排序、数字格式等。

LOCALE 中有一个比较重要的规则 LC_COLLATE，即排序方式（Collation），应用它会对数据库行为有显著影响。例如，默认的 libc 库中 zh_CN 提供了 iso14651_t1_pinyin 排序规则，这是一个基于拼音的排序规则。

下面创建一个包含 7 个汉字的表。

```
create table some_chinese(
    name text primary key
);
```

```
insert into some_chinese values
('阿'),('波'),('磁'),('得'),('饿'),('佛'),('割');
```

当采用 zh_CN 提供的拼音排序规则时,按照拼音排序规则排序而不是按照默认的 ASCII 码排序规则排序会更合适。

```
postgres=# select * from some_chinese order by name collate "zh_CN";
 name
------
 阿
 波
 磁
 得
 饿
 佛
 割
(7 rows)
```

可以看到,按照 zh_CN 提供的拼音排序规则排序得到的结果。当然,这个结果取决于 zh_CN 提供的拼音排序规则的具体定义,而不是数据库本身的定义,数据库本身提供的排序规则就是 C。

PostgreSQL 中使用非 C LOCALE 的负面影响是:特定排序规则对涉及字符串大小比较的操作有巨大的性能影响,同时会导致无法在 LIKE 语句中使用普通索引。

C LOCALE 是由数据库本身确保在任何操作系统与平台上使用的,对 Collation 的支持依赖于操作系统。

在实际环境中,操作系统的 Collation 的版本可能发生变化,在 PostgreSQL 15 中会记录 Collation 的版本,使用 pg_database_collation_actual_version 函数进行查询。以下示例中的 5 是当前 postgres 数据库的对象标识符。

```
postgres=# select pg_database_collation_actual_version(5);
 pg_database_collation_actual_version
--------------------------------------
 2.17
(1 row)
```

在数据库运行过程中 Collation 的版本发生改变时,PostgreSQL 15 支持刷新 Collation 的版本,从而避免数据损坏。

```
postgres=# alter database postgres refresh collation version ;
NOTICE:  version has not changed
ALTER DATABASE
```

PostgreSQL 中除了可以使用默认的 C LOCALE,依赖于操作系统,还可以使用第三方的本地化库,如 ICU 标准库来提供排序规则。在 PostgreSQL 15 之前的版本中,ICU 标准库不支持单独的 Collate 和 Ctype 设置,只能先创建一个 Collation,再在 Collation 中指定 ICU Provider。

```
create collation chinese (provider = icu, locale = 'zh');
```

此外,还可以设置 colCaseFirst 选项。

```
create collation upperfirst (provider = icu, locale = 'en@colCaseFirst=upper');
```

colCaseFirst 选项可以对指定表的列使用,但并不能对整个数据库使用。

在 PostgreSQL 15 中,可以在初始化数据库实例时设置默认的全局 ICU 标准库。

```
initdb -D pgdata --locale-provider=icu --icu-locale=en
```

同样，可以设置 colCaseFirst 选项。

```
initdb -D pgdata --locale-provider=icu --icu-locale='en@colCaseFirst=upper'
```

如果在初始化数据库实例时没有设置 colCaseFirst 选项，那么在新建数据库时也可以设置 colCaseFirst 选项。

```
create database db_icu
icu_locale ='en@colCaseFirst=upper'
locale_provider ='icu'
template=template0;
```

要使用 ICU 标准库的特性就需要数据库在编译时打开 --with-icu 选项，默认数据库在编译时关闭这个选项。

6.2.4 流复制支持 UNLOGGED 序列

在 PostgreSQL 15 之前的版本中使用 UNLOGGED 属性创建表可以提升性能，但不适用于主备流复制场景，因为流复制依赖于 WAL。为了配合在流复制场景下 UNLOGGED 表的功能，PostgreSQL 15 新引入了 UNLOGGED 序列来排除复制，以提高流复制的可靠性。

在使用自增序列或标识列创建表时，LOGGED/UNLOGGED 属性会同步继承到标识列或自增序列中。标识列同步继承 UNLOGGED 属性的示例如下。

```
postgres=# create unlogged table test1(
          id bigint generated by default as identity primary key
       );
CREATE TABLE

postgres=# \d test1
               Unlogged table "public.test1"
 Column |  Type  | Collation | Nullable |             Default
--------+--------+-----------+----------+----------------------------------
 id     | bigint |           | not null | generated by default as identity
Indexes:
    "test1_pkey" PRIMARY KEY, btree (id)

postgres=# \d test1_id_seq
             Unlogged sequence "public.test1_id_seq"
  Type  | Start | Minimum |       Maximum       | Increment | Cycles? | Cache
--------+-------+---------+---------------------+-----------+---------+-------
 bigint |   1   |    1    | 9223372036854775807 |     1     | no      |   1
Sequence for identity column: public.test1.id
```

自增序列同步继承 UNLOGGED 属性的示例如下。

```
postgres=# create unlogged table test2(
          id bigserial primary key
       );
CREATE TABLE

postgres=# \d test2
              Unlogged table "public.test2"
```

```
 Column |  Type   | Collation | Nullable |             Default
--------+---------+-----------+----------+---------------------------------
 id     | bigint  |           | not null |
   nextval('test2_id_seq'::regclass)
Indexes:
    "test2_pkey" PRIMARY KEY, btree (id)

postgres=# \d test2_id_seq
             Unlogged sequence "public.test2_id_seq"
  Type  | Start | Minimum |       Maximum       | Increment | Cycles? | Cache
--------+-------+---------+---------------------+-----------+---------+-------
 bigint |   1   |    1    | 9223372036854775807 |     1     |   no    |   1
Owned by: public.test2.id
```

注意，用户可以根据需要单独修改这个持久化属性，使用 ALTER SEQUENCE SET LOGGED/UNLOGGED 语句可以随时修改。

6.2.5　pg_rewind 工具指定外部配置文件

PostgreSQL 11 改进了 pg_rewind 工具对超级用户权限的依赖。PostgreSQL 13 对流复制恢复配置及恢复自动崩溃方面进行了改进。在 PostgreSQL 14 允许使用 pg_rewind 工具通过备节点同步数据来减轻主节点的压力。在 PostgreSQL 15 中，pg_rewind 工具增加了一个比较有用的选项，即 --config-file 选项，用来指定 postgresql.conf 文件的位置。有些 Linux 的发行版本将 PostgreSQL 的配置文件放到 PGDATA 以外，还有些 HA 软件采用类似方式进行配置管理，使用 --config-file 选项可以提高主备环境操作的可靠性。

6.3　PostgreSQL 15 的运维管理优化

PostgreSQL 15 的运维管理优化主要体现在服务端本地备份引入、JSON 格式日志引入、创建数据库功能增强、COPY 操作对文本格式增强、执行计划显示信息改进、pg_receivewal 工具压缩功能增强、PSQL 工具优化。

6.3.1　服务端本地备份引入

在 PostgreSQL 15 之前的版本中备份数据时，只能将数据备份到运行 pg_basebackup 工具的主机上，而 PostgreSQL 15 对 pg_basebackup 工具增加了 target 选项，可以指定多种不同的目标位置。

target 选项有如下几种设置。
- target=client：默认设置，将数据备份到 pg_basebackup 工具的主机上。
- target=blackhole：丢弃备份数据，可以用于测试或调试场景。
- target=server：备份数据文件可写入服务端本地文件系统而无须发送到客户端。

- target=shell：可以通过 basebackup_to_shell 模块进行备份，也可以通过第三方扩展插件备份至云端存储。

要在服务端备份就需要用户具有流复制权限和服务端写文件权限，以 dba 数据库用户为例，权限配置如下。

```
alter user dba replication ;
```

在上面的示例中给用户设置了 replication 权限，否则在备份时会出现如下错误提示。

```
FATAL:  must be superuser or replication role to start walsender
```

设置服务端写文件权限。

```
grant pg_write_server_files to dba;
```

如果不设置 pg_write_server_files，那么在备份时会出现如下错误提示。

```
ERROR:  must be superuser or a role with privileges of the pg_write_server_files
  role to create backup stored on server
```

dba 数据库用户可以发起服务端本地备份，同时可以结合新增的压缩方法使用，代码如下。

```
$ pg_basebackup -h x.x.x.x -p x -U dba \
--progress --verbose \
--max-rate=1M \
--checkpoint=fast \
--wal-method=fetch \
--target=server:/tmp/backup \
--compress=server-lz4:level=5
```

上述代码的第二行中的选项用于观察备份过程和进度；第三行中的--max-rate 选项用于对备份数据进行限速；第四行中的--checkpoint 选项的值如果不为 fast，那么会使用默认的 spread。基础备份需要伴随一次 CHECKPOINT 操作，spread 模式会等待定期的 CHECKPOINT 操作触发完成，这可能需要较长的时间，但对运行的系统影响较小。在用户想要快速完成备份时，可以使用 fast 模式立即执行 CHECKPOINT 操作，以消耗更多的 I/O 为代价来快速完成备份；第五行中的--wal-method 选项的值不能为默认的 stream，这是因为要获取 WAL 就需要使用 none 模式或 fetch 模式；第六行中的--compress 选项的设置可以参考 6.1.5 节中的内容。

上述代码的执行结果如下。

```
pg_basebackup: initiating base backup, waiting for checkpoint to complete
pg_basebackup: checkpoint completed
pg_basebackup: write-ahead log start point: 0/5E000028 on timeline 1
1600041/1600041 kB (100%), 1/1 tablespace
pg_basebackup: write-ahead log end point: 0/5E000138
pg_basebackup: base backup completed
```

由于在备份时采用了 lz4 算法进行压缩，因此在恢复时需要先进行如下解压缩操作。

```
$ cd /tmp/backup/

$ lz4 -d base.tar.lz4

$ tar -xvf base.tar
```

然后进行备份恢复。

6.3.2 JSON 格式日志引入

PostgreSQL 15 之前的版本中的数据库日志支持 LOG 和 CSV 两种格式，PostgreSQL 15 新增了 JSON 格式。JSON 格式日志非常方便被一些工具进行转储和分析。

下面演示 JSON 格式日志在 PostgreSQL 15 中的使用过程。

设置如下参数。

```
postgres=# show logging_collector;
 logging_collector
-------------------
 on
(1 row)

postgres=# show log_destination;
 log_destination
-----------------
 jsonlog
(1 row)
```

与使用 CSV 格式日志类似，参数 log_filename 并不需要把后缀 log 修改为 json。参数 log_filename 的配置如下。

```
postgres=# show log_filename;
 log_filename
---------------
 pg_log_%u.log
(1 row)
```

设置参数 log_destination 的值为 jsonlog 后，重载参数，就可以观察到 *.json 文件了。

```
$ ls -l /opt/pgdata1500/log/
total 590316
-rw------- 1 postgres dba      545218 May 23 18:37 pg_log_1.json
-rw------- 1 postgres dba         336 May 23 15:42 pg_log_1.log
-rw------- 1 postgres dba       21869 May 24 14:01 pg_log_2.json
-rw------- 1 postgres dba         167 May 24 07:28 pg_log_2.log
-rw------- 1 postgres dba       13363 May 25 22:09 pg_log_3.json
-rw------- 1 postgres dba         167 May 25 09:33 pg_log_3.log
-rw------- 1 postgres dba       99502 May 26 21:00 pg_log_4.json
-rw------- 1 postgres dba         668 May 26 15:10 pg_log_4.log
-rw------- 1 postgres dba       97839 May 20 18:01 pg_log_5.json
-rw------- 1 postgres dba         835 May 20 17:35 pg_log_5.log
-rw------- 1 postgres dba      365840 Jun  4 09:07 pg_log_6.json
-rw------- 1 postgres dba   603164236 Jun  4 08:44 pg_log_6.log
-rw------- 1 postgres dba      134464 May 22 20:46 pg_log_7.json
-rw------- 1 postgres dba         835 May 22 12:29 pg_log_7.log
```

JSON 格式日志相比 CSV 格式日志主要有以下两个优势：
- 可以正确处理日志跨行问题。
- 可以保持干净且有效的日志输出项。

针对第一个优势，在使用 CSV 格式日志时，如果语句比较复杂且内容很长需占用多行，那么日志将不能正确解析。JSON 格式日志保留了语句的换行符，可以观察下面的示例中 JSON 格式日志的 message 属性。

```
{
"timestamp":"2022-04-20 10:36:26.125 CST",
"user":"postgres",
"dbname":"postgres",
"pid":3881,
"remote_host":"[local]",
"session_id":"625f71aa.f29",
"line_num":1,
"ps":"idle",
"session_start":"2022-04-20 10:36:26 CST",
"vxid":"3/2",
"txid":0,
"error_severity":"LOG",
"message":"statement: select\n    relname,\n    relkind\nfrom\n    pg_class\nlimit
   1;",
"application_name":"psql",
"backend_type":"client backend",
"query_id":0
}
```

针对第二个优势，在使用 CSV 格式日志时，日志项是固定的，在系统后台进程与用户后端进程的日志项中并不全都填满了内容，CSV 格式日志中有很多日志空项。JSON 格式日志对这一问题进行了优化，精简了无实际内容的日志项，使得每行日志更加干净、清爽。下面的一条系统后台进程 CHECKPOINT 操作日志输出得非常简洁。

```
{
"timestamp":"2022-04-20 10:24:47.710 CST",
"pid":3383,
"session_id":"625f67e3.d37",
"line_num":12,
"session_start":"2022-04-20 09:54:43 CST",
"txid":0,
"error_severity":"LOG",
"message":"checkpoint complete: wrote 4 buffers (0.0%); 0 WAL file(s) added, 0
   removed, 0 recycled; write=0.303 s, sync=0.135 s, total=0.809 s; sync files=4,
   longest=0.127 s, average=0.034 s; distance=0 kB, estimate=114 kB",
"backend_type":"checkpointer",
"query_id":0
}
```

除此之外，数据库日志也可以使用 file_fdw 模块进行映射，并使用 SQL 语句进行分析。下面演示配置的使用过程。

创建 file_fdw 模块。

```
create extension file_fdw;
```

创建服务器。

```
create server pglog foreign data wrapper file_fdw;
```

创建 CSV 格式日志的映射表。

```
CREATE FOREIGN TABLE public.pg_csvlog_1(
  log_time timestamp(3) with time zone,
```

```
    user_name text,
    database_name text,
    process_id integer,
    connection_from text,
    session_id text,
    session_line_num bigint,
    command_tag text,
    session_start_time timestamp with time zone,
    virtual_transaction_id text,
    transaction_id bigint,
    error_severity text,
    sql_state_code text,
    message text,
    detail text,
    hint text,
    internal_query text,
    internal_query_pos integer,
    context text,
    query text,
    query_pos integer,
    location text,
    application_name text,
    backend_type text,
    leader_pid integer,
    query_id bigint
) SERVER pglog
OPTIONS ( filename '/opt/pgdata1405/log/pg_log_1.csv', header 'true', format
    'csv' );
```

创建 JSON 格式日志的映射表。

```
CREATE FOREIGN TABLE pg_jsonlog_5(
jsonstr text
 ) SERVER local_file_server
OPTIONS (program 'cat /opt/pgdata1500/log/pg_log_5.json |jq -cMR');
```

由于配置了日志按周每天循环覆盖，因此应按单个日志文件进行映射，其他日志文件进行映射的配置类似。JSON 日志进行映射解析借助了 jq 工具，使用 jp 工具可以非常方便地读取 JSON 日志的原始内容。

下面将整行原始文本使用 JSON 构造函数转换为 JSON 数据类型，并通过创建物化视图来突破外部表不能创建索引的限制。如果日志量变多，那么可以通过对物化视图创建索引来挖掘日志，并进行高效分析。

```
create materialized view mv_pg_jsonlog_5
as select json(trim(jsonstr,'"')) as jsonlog from pg_jsonlog_5;
```

可以在客户端通过 SQL 接口分析服务端的日志。

如果使用不同的工具对日志有不同的格式需求，那么可以将参数 log_destination 设置为多种格式。例如，针对下面的设置，同样的日志内容将输出 3 种不同的格式。

```
$ grep log_destination /opt/pgdata1500/postgresql.conf
log_destination = 'stderr,csvlog,jsonlog'
```

6.3.3 创建数据库功能增强

在 PostgreSQL 15 中创建数据库时，新增了一些选项。在使用 createdb 工具创建数据库时，新增了 3 个选项：--icu-locale、--locale-provider、--strategy，在使用 create database 工具创建数据库时，新增了 2 个选项：collation_version 和 oid。本节主要介绍 oid 选项和 strategy 选项。

实例在初始化后，默认包含 3 个数据库：template1、template0 和 postgres。这 3 个数据库的 oid 选项的值在 PostgreSQL 15 中变成了更简单的数字，即 1、4、5。这 3 个值代表了创建数据库的顺序。首先创建 1 号数据库 template1，其次创建 4 号数据库 template0，最后创建 5 号数据库 postgres。

```
postgres=# select oid,datname from pg_database ;
 oid | datname
-----+-----------
   5 | postgres
   1 | template1
   4 | template0
(3 rows)
```

oid 选项的值在 CREATE USER 时也可以自定义，这样比较有意义。可以固化为一个数字，这样当查看数据文件及进行一些关联查询时会比较方便。例如，创建一个 oid 选项的值为 888888 的数据库。

```
postgres=# create database foo oid 888888;
CREATE DATABASE
```

在自定义数据库的 oid 选项的值时，需要注意系统的保留设置范围：不能小于 2 的 14 次方（16384）。

strategy 选项有如下两个值。
- wal_log：默认值，对模板数据库的数据文件逐个复制数据块并写入 WAL。这种方式不会触发 CHECKPOINT 操作。当模板数据库不大时，创建效率非常高。
- file_copy：保持与 PostgreSQL 15 之前的版本一致的行为，记录 WAL 的量很少，会强制触发 CHECKPOINT 操作。当模板数据库较大时，可以显著降低 WAL 的写入量。

要保持 PostgreSQL 15 之前的版本的兼容性，需要设置 strategy 选项的值为 file_copy。

```
create database foo1 with strategy='file_copy';
```

6.3.4 COPY 操作对文本格式增强

对单表进行数据与文件之间的复制通常使用元命令\copy 在客户端进行导入和导出，在处理较长的复杂语句生成的结果集时，会面临语句跨行的问题。此时可以使用临时表先存储数据再进行导入和导出，或编辑多行语句，将其调整到一行。除此以外，还可以通过服务端 COPY 进行处理。

```
copy (
  select *
  from foo
```

```
       where a = 'xx'
) to stdout with csv header \g foo.csv
```

在上面的示例中，先通过服务端 COPY 输出到标准输出中，再使用元命令\g 将标准输出写入本地文件。

PostgreSQL 15 支持使用 header 选项把表的列名导出到文件头，同时在导入数据时还可以对导出部分列的数据使用 header 选项的 match 值匹配顺序。

下面通过示例进行演示。

创建测试表。

```
create table foo(a text, b text, c text);
insert into foo select 'A', 'B', 'C';
```

PostgreSQL 15 之前的版本仅支持 CSV 格式使用 header 选项，使用文本格式会提示错误。

```
postgres=# copy foo (a,b,c) to '/tmp/foo.txt' with (format 'text',header on);
ERROR:  COPY HEADER available only in CSV mode
```

在下面的代码中，小写的 a、b、c 为 foo 表的列名，大写的 A、B、C 分别是对应列的数据。

```
postgres=# copy foo (a,b,c) to '/tmp/foo.txt' with (format 'text',header on);
COPY 1
postgres=# \! cat /tmp/foo.txt
a       b       c
A       B       C
```

另外，也可以对部分列的数据按指定顺序进行导出。

```
postgres=# copy foo (c,b) to '/tmp/foo.txt' with (format 'text',header on);
COPY 1
postgres=# \! cat /tmp/foo.txt
c       b
C       B
```

在导入数据时，使用 header 选项的 match 值对导入文件首行的列名顺序进行匹配检测。由于下面导入列名的指定顺序是 b、c，而在上一步中导出列名的指定顺序是 c、b，因此在导入数据时会有如下错误提示。

```
postgres=# copy foo (b,c) from '/tmp/foo.txt' with (format 'text',header match);
ERROR:  column name mismatch in header line field 1: got "c", expected "b"
CONTEXT:  COPY foo, line 1: "c       b"
```

此外，file_fdw 模块也增加了文本格式的 header 选项，示例如下。

```
create extension file_fdw;
create server file_server foreign data wrapper file_fdw;
create foreign table f_foo(a text, b text, c text)
    server file_server options (
filename '/tmp/foo.txt',
format 'text',
header 'true');
```

6.3.5 执行计划显示信息改进

要在 PostgreSQL 中查看执行计划应使用 EXPLAIN 命令，虽然执行计划显示的信息对开发人员并不特别友好，但 PostgreSQL 可以理解执行计划，执行计划包含表的行数、去重值、常用值等信息。通过执行计划可以看到 SQL 语句执行的规划路线，以及各执行节点的资源和时间消耗。在优化某条 SQL 语句或排查 SQL 语句执行的问题时，通常应先使用 EXPLAIN 命令查看执行计划。

在 PostgreSQL 15 中，执行计划的显示信息有两处改进。
- 在 verbose 模式下，固化显示临时表对象的模式名称。
- 在 buffers 模式下，显示临时文件块读写的耗时信息。

第一处改进是在 verbose 模式下，固化显示临时表对象的模式名称。在特殊模式下，临时表的实际模式名称是动态变化的。在下面的代码中，后缀 6 是一个动态变化的 BackendID，表示本进程在内存进程数组中的序号，标识当前连接是哪个会话。

```
postgres=# create temp table tmp();
CREATE TABLE
postgres=# \d tmp
          Table "pg_temp_6.tmp"
 Column | Type | Collation | Nullable | Default
--------+------+-----------+----------+---------
```

在 PostgreSQL 15 之前的版本中使用 verbose 模式查看执行计划时，动态显示临时表对象的模式名称，示例如下。

```
postgres=# create temp table tmp1(id int);
CREATE TABLE
postgres=# explain verbose select * from tmp1;
                      QUERY PLAN
-----------------------------------------------------------------
 Seq Scan on pg_temp_6.tmp1  (cost=0.00..35.50 rows=2550 width=4)
   Output: id
(2 rows)
```

在 PostgreSQL 15 中使用 verbose 模式查看执行计划时，固化显示临时表对象的模式名称，示例如下。

```
postgres=# explain verbose select * from tmp1;
                      QUERY PLAN
-----------------------------------------------------------------
 Seq Scan on pg_temp.tmp1  (cost=0.00..7856.15 rows=564315 width=4)
   Output: id
(2 rows)
```

第二处改进是在 buffers 模式下，显示临时文件块读写的耗时信息。需要先打开参数 track_io_timing，这样执行计划才可以在 buffers 模式下，通过 I/O Timings 显示临时文件块读写的耗时信息，示例如下。

```
postgres=# explain (buffers,costs off,analyze)
             insert into tmp1 select * from generate_series(1,100000);
                      QUERY PLAN
-----------------------------------------------------------------
 Insert on tmp1 (actual time=164.341..164.342 rows=0 loops=1)
   Buffers: local hit=100884 dirtied=442 written=884, temp read=171 written=171
   I/O Timings: temp read=0.612 write=0.977
```

```
        -> Function Scan on generate_series (actual time=38.708..74.118 rows=100000
    loops=1)
            Buffers: temp read=171 written=171
            I/O Timings: temp read=0.612 write=0.977
 Planning Time: 0.065 ms
 Execution Time: 165.008 ms
(8 rows)
```

6.3.6 pg_receivewal 工具压缩功能增强

在 PostgreSQL 中可以使用 pg_receivewal 工具在线备份 WAL，从而起到虚拟备库的作用。使用 pg_receivewal 工具比使用归档命令备份 WAL 更加安全，同时不需要等待 WAL 文件写满。该工具基于流复制协议流式传输 WAL，当数据库服务发生重启时，不会发生数据丢失或数据损坏的情况，该工具通常与-Z/--compress 选项搭配使用。

PostgreSQL 15 之前的版本使用 pg_receivewal 工具，只能通过--compress 选项设置 gzip 算法的压缩级别，而 PostgreSQL 15 可以通过--compress 选项指定压缩方法和压缩级别，语法如下。

```
--compress=METHOD[:DETAIL]
```

要指定压缩方法，可以使用 gzip 算法或新增的 lz4 算法。

具体的使用示例如下。

```
$ mkdir /tmp/wal

$ pg_receivewal -D /tmp/wal --compress=lz4:9 -v
pg_receivewal: starting log streaming at 1/55000000 (timeline 1)
```

此时，在/tmp/wal 目录中可以看到如下文件。

```
$ ls /tmp/wal
000000010000000100000055.lz4.partial
```

当数据库服务发生重启时，应先关闭数据库服务，继续查看 pg_receivewal 工具窗口。

```
pg_receivewal: error: replication stream was terminated before stop point
pg_receivewal: disconnected; waiting 5 seconds to try again
pg_receivewal: error: connection to server on socket "/tmp/.s.PGSQL.1501" failed:
    No such file or directory
Is the server running locally and accepting connections on that socket?
...
pg_receivewal: disconnected; waiting 5 seconds to try again
pg_receivewal: starting log streaming at 1/55000000 (timeline 1)
```

从上述代码中可以看出，使用 pg_receivewal 工具能够自动进行流复制重连，不会发生数据丢失的情况。使用 lz4 算法编译时需要打开--with-lz4 选项。

6.3.7 PSQL 工具优化

PostgreSQL 15 对 PSQL 工具主要进行了如下优化。

- 元命令\copy 的性能提升。
- 元命令\lo_list 和\dl 支持使用 "+" 选项查看对象权限。

- 元命令\g 支持将多个查询结果输出到文件中。
- 新增元命令\getenv，支持将操作系统环境变量赋值给 PSQL 工具变量。
- 新增元命令\dconfig，支持查看服务端的非默认参数或模糊匹配多个变量。

下面重点演示元命令\g、\getenv、\dconfig。

元命令\g 的示例如下。

```
postgres=# select 1\; select 2 \g file
postgres=# \! cat file
 ?column?
----------
        1
(1 row)


 ?column?
----------
        2
(1 row)
```

PostgreSQL 15 之前的版本只能输出最后一个查询结果。

对于元命令\getenv，操作系统环境变量 PGHOME 的示例如下。

```
$ env | grep PGHOME
PGHOME=/opt/pg15
```

下面通过元命令\getenv 先将操作系统环境变量 PGHOME 赋值给自定义的 PSQL 工具变量 pg_home，再通过元命令\echo 输出正确的值。

```
postgres=# \getenv pg_home PGHOME
postgres=# \echo :pg_home
/opt/pg15
```

元命令\dconfig 类似于 show 命令，可以查看服务端的多个参数，也支持通配符模糊匹配参数名称，以及使用"+"选项查看参数类型、生效范围及权限信息。例如，查看 AUTOVACUUM 进程的相关参数。

```
postgres=# \dconfig autovacuum*
           List of configuration parameters
             Parameter                      | Value
----------------------------------------+------------
 autovacuum                             | on
 autovacuum_analyze_scale_factor        | 0.1
 autovacuum_analyze_threshold           | 50
 autovacuum_freeze_max_age              | 200000000
 autovacuum_max_workers                 | 3
 autovacuum_multixact_freeze_max_age    | 400000000
 autovacuum_naptime                     | 1min
 autovacuum_vacuum_cost_delay           | 2ms
 autovacuum_vacuum_cost_limit           | -1
 autovacuum_vacuum_insert_scale_factor  | 0.2
 autovacuum_vacuum_insert_threshold     | 1000
 autovacuum_vacuum_scale_factor         | 0.2
 autovacuum_vacuum_threshold            | 50
 autovacuum_work_mem                    | -1
(14 rows)
```

继续查看参数 autovacuum_max_workers 更详细一点的信息，使用"+"选项查看扩展的参数类型、生效范围及权限信息，示例如下：

```
postgres=# \x \dconfig+ autovacuum_max_workers
Expanded display is on.
List of configuration parameters
-[ RECORD 1 ]----+----------------------
Parameter        | autovacuum_max_workers
Value            | 3
Type             | integer
Context          | postmaster
Access privileges|
```

6.4　PostgreSQL 15 的开发易用性提升

PostgreSQL 15 的开发易用性提升主要体现在 MERGE 语句引入、NULL 值与 UNIQUE 约束更搭、numeric 数据类型改进、正则表达式函数引入、分区表改进，以及逻辑复制改进。

6.4.1　MERGE 语句引入

在 SQL Server 及 Oracle 中经常能看到 MERGE 语句，PostgreSQL 15 也引入了 MERGE 语句。使用 MERGE 语句能让需要多行代码的业务逻辑用一条简单的语句实现，不仅可以减少代码的行数，而且可以减少业务逻辑的维护工作量。使用 MERGE 语句可以让 SQL 语句非常便捷地从 Oracle 中迁移到 PostgreSQL 中，同时开发人员的工作也会变得更加容易。

MERGE 语句的使用示例如下：

```
[ WITH with_query [, ...] ]
MERGE INTO target_table_name [ [ AS ] target_alias ]
USING data_source ON join_condition
when_clause [...]

where data_source is:

{ source_table_name | ( source_query ) } [ [ AS ] source_alias ]

and when_clause is:

{ WHEN MATCHED [ AND condition ] THEN { merge_update | merge_delete | DO NOTHING }
  |
  WHEN NOT MATCHED [ AND condition ] THEN { merge_insert | DO NOTHING } }

and merge_insert is:

INSERT [( column_name [, ...] )]
```

```
[ OVERRIDING { SYSTEM | USER } VALUE ]
{ VALUES ( { expression | DEFAULT } [, ...] ) | DEFAULT VALUES }

and merge_update is:

UPDATE SET { column_name = { expression | DEFAULT } |
             ( column_name [, ...] ) = ( { expression | DEFAULT } [, ...] ) } [, ...]

and merge_delete is:

DELETE
```

可以使用单个 MERGE 语句在一个事务中根据源表和目标表的条件匹配对目标表执行 INSERT 操作、UPDATE 操作或 DELETE 操作。

如果满足匹配条件，那么会有 3 种分支。

- 用源表去更新目标表。
- 用源表去删除目标表。
- 什么也不做。

如果不满足匹配条件，那么会有两种分支。

- 用源表去插入目标表。
- 什么也不做。

根据是否匹配及分支组合，MERGE 语句通常有 6 种常用的使用模式，这 6 种常用的使用模式的应用示例如下。

创建 3 个测试表。

```
create table a_merge (
    id   int not null,
    name varchar not null,
    year int
);

create table b_merge (
    id   int not null,
    aid  int not null,
    name varchar not null,
    year int,
    city varchar
);

create table c_merge (
    id   int not null,
    name varchar not null,
    city varchar not null
);
```

- 模式一：若匹配则更新，若不匹配则插入。

向 a_merge 表和 b_merge 表中插入数据。

```
insert into a_merge values(1,'liuwei',20);
insert into a_merge values(2,'zhangbin',21);
```

```
insert into a_merge values(3,'fuguo',20);

insert into b_merge values(1,2,'zhangbin',30,'吉林');
insert into b_merge values(2,4,'yihe',33,'黑龙江');
insert into b_merge (id,aid,name,city) values(3,3,'fuguo','山东');
```

查看 a_merge 表和 b_merge 表中的数据。

```
postgres=# select * from a_merge;select * from b_merge;
 id |   name   | year
----+----------+------
  1 | liuwei   |   20
  2 | zhangbin |   21
  3 | fuguo    |   20
(3 rows)

 id | aid |   name   | year | city
----+-----+----------+------+--------
  1 |   2 | zhangbin |   30 | 吉林
  2 |   4 | yihe     |   33 | 黑龙江
  3 |   3 | fuguo    |      | 山东
(3 rows)
```

使用 b_merge 表更新 a_merge 表中的数据。

```
postgres=# merge into a_merge a
           using (select b.aid,b.name,b.year from b_merge b) c
              on (a.id=c.aid)
           when matched then
              update set year=c.year
           when not matched then
              insert values(c.aid,c.name,c.year);
MERGE 3
```

查看 a_merge 表中的数据。

```
postgres=# select * from a_merge;
 id |   name   | year
----+----------+------
  1 | liuwei   |   20
  2 | zhangbin |   30
  3 | fuguo    |
  4 | yihe     |   33
(4 rows)
```

从执行结果中可以看出，b_merge 表中的两条匹配的数据更新到了 a_merge 表中，b_merge 表中的一条未匹配的数据插入到了 a_merge 表中。

- 模式二：仅匹配更新。

向 b_merge 表中插入两条数据，一条语句插入的值，源表可以匹配目标表；另一条语句插入的值，源表不可以匹配目标表。

```
insert into b_merge values(4,1,'liuwei',80,'江西');
insert into b_merge values(5,5,'tiantian',23,'河南');
```

查看 a_merge 表和 b_merge 表中的数据。
```
postgres=# select * from a_merge;select * from b_merge;
```

```
 id |   name   | year
----+----------+------
  1 | liuwei   |   20
  2 | zhangbin |   30
  3 | fuguo    |
  4 | yihe     |   33
(4 rows)


 id | aid |   name   | year | city
----+-----+----------+------+--------
  1 |   2 | zhangbin |   30 | 吉林
  2 |   4 | yihe     |   33 | 黑龙江
  3 |   3 | fuguo    |      | 山东
  4 |   1 | liuwei   |   80 | 江西
  5 |   5 | tiantian |   23 | 河南
(5 rows)
```

使用 b_merge 表更新 a_merge 表中的数据，注意仅有 UPDATE 语句。

```
postgres=# merge into a_merge a
           using (select b.aid,b.name,b.year from b_merge b) c
              on (a.id=c.aid)
        when matched then
          update set year=c.year;
MERGE 4
```

MERGE 语句执行完成后，查看 a_merge 表中的数据。

```
postgres=# select * from a_merge;
 id |   name   | year
----+----------+------
  1 | liuwei   |   80
  2 | zhangbin |   30
  3 | fuguo    |
  4 | yihe     |   33
(4 rows)
```

从执行结果中可以看出，仅对 b_merge 表中匹配的数据更新了 a_merge 表。

- 模式三：仅不匹配插入。

改变 b_merge 表中的一条数据进行测试。

```
update b_merge set year=70 where aid=2;
```

查看 a_merge 表和 b_merge 表中的数据。

```
postgres=# select * from a_merge;select * from b_merge;
 id |   name   | year
----+----------+------
  1 | liuwei   |   80
  2 | zhangbin |   30
  3 | fuguo    |
  4 | yihe     |   33
(4 rows)

 id | aid |   name   | year | city
----+-----+----------+------+--------
```

```
 2 |  4 | yihe     | 33 | 黑龙江
 3 |  3 | fuguo    |    | 山东
 4 |  1 | liuwei   | 80 | 江西
 5 |  5 | tiantian | 23 | 河南
 1 |  2 | zhangbin | 70 | 吉林
(5 rows)
```

使用 b_merge 表更新 a_merge 表中的数据,注意仅有 INSERT 语句。

```
postgres=# merge into a_merge a
           using (select b.aid,b.name,b.year from b_merge b) c
             on (a.id=c.aid)
         when not matched then
             insert values(c.aid,c.name,c.year);
MERGE 1
```

MERGE 语句执行完成后,查看 a_merge 表中的数据。

```
postgres=# select * from a_merge;
 id | name     | year
----+----------+------
  1 | liuwei   | 80
  2 | zhangbin | 30
  3 | fuguo    |
  4 | yihe     | 33
  5 | tiantian | 23
(5 rows)
```

从执行结果中可以看出,仅对 b_merge 表中不匹配的数据插入了 a_merge 表。

- 模式四:二次匹配。

当在 ON 条件中进行条件匹配之后,还可以在其后的 when 子句中对 ON 条件筛选出来的数据再次进行条件判断,用来控制哪些数据要更新,哪些数据要插入,哪些数据要删除。

调整 b_merge 表中的数据。

```
update b_merge set name='yihe++' where id=2;
update b_merge set name='liuwei++' where id=4;
insert into b_merge values(6,6,'ningqin',23,'江西');
insert into b_merge values(7,7,'bing',24,'四川');
```

查看 a_merge 表和 b_merge 表中的数据。

```
postgres=# select * from a_merge;select * from b_merge;
 id | name     | year
----+----------+------
  1 | liuwei   | 80
  2 | zhangbin | 30
  3 | fuguo    |
  4 | yihe     | 33
  5 | tiantian | 23
(5 rows)

 id | aid | name     | year | city
----+-----+----------+------+--------
  3 |  3  | fuguo    |      | 山东
```

```
 5 |   5 | tiantian |  23 | 河南
 1 |   2 | zhangbin |  70 | 吉林
 2 |   4 | yihe++   |  33 | 黑龙江
 4 |   1 | liuwei++ |  80 | 江西
 6 |   6 | ningqin  |  23 | 江西
 7 |   7 | bing     |  24 | 四川
(7 rows)
```

使用 b_merge 表更新 a_merge 表中的数据，注意应分别在执行 INSERT 操作和 UPDATE 操作之前添加二次匹配的条件限制，控制数据的插入和更新。

```
postgres=# merge into a_merge a
           using (select b.aid,b.name,b.year,b.city from b_merge b) c
              on (a.id=c.aid)
           when matched and c.city != '江西' then
              update set name=c.name
           when not matched and c.city = '江西' then
              insert values(c.aid,c.name,c.year);
MERGE 5
```

MERGE 语句执行完成后，查看 a_merge 表中的数据。

```
postgres=# select * from a_merge ;
 id |   name   | year
----+----------+------
  1 | liuwei   |  80
  2 | zhangbin |  30
  3 | fuguo    |
  4 | yihe++   |  33
  5 | tiantian |  23
  6 | ningqin  |  23
(6 rows)
```

从执行结果中可以看出，匹配和不匹配分支都分别进行了二次匹配，符合预期。

- 模式五：无条件全部插入。

下面期望将 b_merge 表中的数据无条件地全部插入到 c_merge 表中，MERGE 语句可以将 ON 条件设置为假，使用不匹配的分支插入。

```
postgres=# merge into c_merge c
           using (select b.aid,b.name,b.city from b_merge b) b
              on (1=0)
           when not matched then
              insert values(b.aid,b.name,b.city);
MERGE 7
```

MERGE 语句执行完成后，查看 c_merge 表中的数据。

```
postgres=# select * from c_merge ;
 id |   name   |  city
----+----------+--------
  3 | fuguo    | 山东
  5 | tiantian | 河南
  2 | zhangbin | 吉林
  4 | yihe++   | 黑龙江
  1 | liuwei++ | 江西
  6 | ningqin  | 江西
```

```
 7 | bing      | 四川
(7 rows)
```

从执行结果中可以看出，b_merge 表中的数据全部插入到了 c_merge 表中。
- 模式六：仅匹配删除。

在 when 子句中除了可以执行 UPDATE 操作，也可以执行 DELETE 操作。查看 a_merge 表和 b_merge 表中的数据。

```
postgres=# select * from a_merge;
 id |   name   | year
----+----------+------
  1 | liuwei   |   80
  2 | zhangbin |   30
  3 | fuguo    |
  4 | yihe++   |   33
  5 | tiantian |   23
  6 | ningqin  |   23
(6 rows)

postgres=# select * from b_merge;
 id | aid |   name   | year | city
----+-----+----------+------+--------
  3 |   3 | fuguo    |      | 山东
  5 |   5 | tiantian |   23 | 河南
  1 |   2 | zhangbin |   70 | 吉林
  2 |   4 | yihe++   |   33 | 黑龙江
  4 |   1 | liuwei++ |   80 | 江西
  6 |   6 | ningqin  |   23 | 江西
  7 |   7 | bing     |   24 | 四川
(7 rows)
```

使用 b_merge 表来匹配删除 a_merge 表，同时在删除时进行二次匹配，只删除江西的数据。

```
postgres=# merge into a_merge a
           using (select b.aid,b.name,b.year,b.city from b_merge b) c
              on (a.id=c.aid)
        when matched and c.city = '江西' then
            delete;
MERGE 2
```

MERGE 语句执行完成后，查看 a_merge 表中的数据。

```
postgres=# select * from a_merge;
 id |   name   | year
----+----------+------
  2 | zhangbin |   30
  3 | fuguo    |
  4 | yihe++   |   33
  5 | tiantian |   23
(4 rows)
```

从执行结果中可以看出，只有符合匹配条件且二次匹配条件满足江西的两条数据被删除了。注意，MERGE 语句中的 DELETE 操作不需要任何参数，这是因为 MERGE 语句中需要被删除的行或列的信息已经非常清晰了。

6.4.2　NULL 值与 UNIQUE 约束更搭

　　约束是定义在表字段中的规则，可以在创建表的语句中添加字段约束或表约束（可以跨多个字段），也可以在创建表之后通过 ALTER TABLE 命令添加约束。如果用户在存储字段数据时违反了约束，那么会抛出错误提示。

　　UNIQUE 约束可以保证在一个字段或一组字段中的数据相较于表中其他行的数据是唯一的。在 PostgreSQL 中，需要注意 NULL 值，SQL 标准允许 UNIQUE 约束有多个 NULL 值，由于 NULL 值是未知的，很难判定一个未知的值与另外一个未知的值是否相等，因此 NULL 值并不会违反 UNIQUE 约束的规则。

　　PostgreSQL 遵守 SQL 标准，UNIQUE 约束允许有多个 NULL 值。

　　下面通过示例进行演示。

　　创建测试表。

```
create table test_null_old(
    id bigint generated by default as identity primary key,
    val1 text not null,
    val2 text null,
    constraint uq_val1_val2 unique(val1, val2)
);
```

　　在两条语句中插入相同的 val2，第二次插入时会失败，这符合预期。

```
postgres=# insert into test_null_old(val1,val2) values('hello', 'world');
INSERT 0 1

postgres=# insert into test_null_old(val1,val2) values('hello', 'world');
ERROR:  duplicate key value violates unique constraint "uq_val1_val2"
DETAIL:  Key (val1, val2)=(hello, world) already exists.
```

　　如果插入的 val2 为 NULL 值，那么可以多次插入。

```
postgres=# insert into test_null_old(val1,val2)
            values('hello', NULL) returning *;
 id | val1  | val2
----+-------+------
  3 | hello |
(1 row)
INSERT 0 1

postgres=# insert into test_null_old(val1,val2)
            values('hello', NULL) returning *;
 id | val1  | val2
----+-------+------
  4 | hello |
(1 row)
INSERT 0 1
```

　　前面已经介绍了 NULL 值可以多次插入的原因，但仍有些开发人员认为重复插入 NULL 值并不符合预期。如果要限制 NULL 值不重复，那么需要使用其他方式进行处理，如使用部分表达式索引。

在 PostgreSQL 15 中，对 UNIQUE 约束新增了 unique nulls not distinct 选项，可以限制 NULL 值不重复。下面继续通过示例进行演示。

创建测试表。

```
create table test_null_new(
   id bigint generated by default as identity primary key,
   val1 text not null,
   val2 text null,
   constraint uq_val1_val2_new
      unique nulls not distinct (val1, val2)
);
```

可以看出，新创建的测试表使用了 unique nulls not distinct 选项。下面在第二次插入 val1 为 hello，val2 为 NULL 值时，会提示违反唯一约束。

```
postgres=# insert into test_null_new(val1,val2)
           values('hello', NULL) returning *;
 id | val1  | val2
----+-------+------
  1 | hello |
(1 row)
INSERT 0 1

postgres=# insert into test_null_new(val1,val2)
           values('hello', NULL) returning *;
ERROR:  duplicate key value violates unique constraint "uq_val1_val2_new"
DETAIL:  Key (val1, val2)=(hello, null) already exists.
```

如果 val1 更换一个值，val2 也被允许为 NULL 值。

```
postgres=# insert into test_null_new(val1,val2)
           values('world', NULL) returning *;
 id | val1  | val2
----+-------+------
  3 | world |
(1 row)
INSERT 0 1
```

PostgreSQL 15 的这个新特性允许开发人员自由控制数据中的 NULL 值是否重复，使得 NULL 值与 UNIQUE 约束更加搭配，同时 NULL 值的默认行为也会继续与 PostgreSQL 15 之前的版本的对应行为保持一致。

6.4.3　numeric 数据类型改进

PostgreSQL 的数据类型有 3 种：分别是整数类型，包括 smallint、integer 和 bigint；精度类型，包括 numeric 和 decimal（其中 decimal 等同于 numeric）；浮点数类型，包括 real 和 double。这 3 种数据类型从 Oracle 迁移到 PostgreSQL 将 number(precision,scale) 转化为 numeric(precision,scale) 时，有如下两种情况需要进行特殊处理。

- scale 小于 0。
- scale 大于 0，并且 precision 小于 scale。

在 PostgreSQL 15 之前的版本中，由于 scale 必须大于或等于 0 且不能超过 precision，因此数据在迁移过来时需要修改。

PostgreSQL 15 允许 numeric 数据类型的 scale 是负数或比 precision 大，同时扩展了 scale 的范围，scale 的范围为[-1000, 1000]，允许 scale 被独立设置而不依赖于精度，precision 的范围仍然为[1，1000]。

scale 为负数表示可以向小数点左侧取整。例如，id 存储范围为绝对值在 10^8 以内的整数，因为 scale 为-3 所以向千位四舍五入取整。

```
postgres=# create table t1(id numeric(5,-3));
CREATE TABLE

postgres=# insert into t1 values(10000512) returning id;
    id
----------
 10001000
(1 row)
INSERT 0 1


postgres=# insert into t1 values(-10000512) returning id;
    id
----------
 -10001000
(1 row)
INSERT 0 1
```

若 scale 比 precision 大则表示分数或小数。例如，id 存储范围为绝对值在 10^{-1} 以内的分数，因为 scale 为 3 所以向千分位四舍五入取整。

```
postgres=# create table t2(id numeric(2,3));
CREATE TABLE

postgres=# insert into t2 values(0.0985) returning id;
   id
-------
 0.099
(1 row)
INSERT 0 1

postgres=# insert into t2 values(-0.0985) returning id;
   id
--------
 -0.099
(1 row)
INSERT 0 1
```

6.4.4 正则表达式函数引入

截至 PostgreSQL 14，共有 6 种函数支持 POSIX 风格的正则表达式。

- substring(string text FROM pattern text)：

提取匹配正则表达式的第一个子串。
- regexp_match(string text, pattern text [, flags text])：

以字符串数组的形式返回字符串与正则表达式匹配的第一个子串。
- regexp_matches(string text, pattern text [, flags text])：

以字符串数组集合的形式返回字符串与正则表达式匹配的第一个子串，第三个可选参数若使用 g 标志则全部匹配。
- regexp_replace(string text, pattern text, replacement text [, flags text])：

替换字符串中第一个正则表达式匹配的子串，第四个可选参数若使用 g 标志则全部替换。
- regexp_split_to_array(string text, pattern text [, flags text])：

使用正则表达式将字符串拆解为字符串数组。
- regexp_split_to_table(string text, pattern text [, flags text])：

使用正则表达式将字符串拆解为字符串集合。

要统计字符串匹配某个子串的次数，以及具体的匹配位置，或要从某个指定的位置开始匹配，使用 PostgreSQL 15 之前的版本很难实现。虽然可以通过编写函数来实现这些功能，但其可行性会变得很差。

PostgreSQL 15 新引入一些原生的正则表达式函数。
- regexp_count(string text, pattern text [, start integer [, flags text]])：

统计字符串匹配某个子串的次数，第三个可选参数用于指定从第 N 个字符开始，第四个可选参数用于设置一些标志位。
- regexp_instr(string text, pattern text [, start integer [, N integer [, endoption integer [, flags text [, subexpr integer]]]]])：

返回正则表达式第 N 次匹配的位置。
- regexp_like(string text, pattern text [, flags text])：

检测字符串是否模糊匹配正则表达式，第三个可选参数若使用 i 标志则可以忽略大小写。
- regexp_substr(string text, pattern text [, start integer [, N integer [, flags text [, subexpr integer]]]])：

返回正则表达式第 N 次匹配的子串。

6.4.5 分区表改进

PostgreSQL 15 对分区表进行了如下改进。
- 在进行分区表查询时如果只扫描少量的分区，那么生成执行计划的时间有改进。
- 默认分区或列表分区键包含多个值，能使用预排序扫描，以避免进行排序操作。
- 分区表允许执行 CLUSTER 命令进行数据重排。
- 跨分区外键 UPDATE 操作有改进：在 PostgreSQL 15 之前的版本中进行 UPDATE 操作是通过在源分区上执行 DELETE 操作且在目标分区上执行 INSERT 操作来完成的，而 PostgreSQL 15 直接在根分区上执行 UPDATE 操作，语义更加清晰。

6.4.6 逻辑复制改进

PostgreSQL 15 对逻辑复制进行了如下改进。

1．支持两阶段提交

PostgreSQL 14 中的逻辑复制只支持对准备好的事务进行逻辑解码，并发送到输出插件中，而订阅端的工作尚未完成。PostgreSQL 15 订阅端支持通过 two_phase 选项开启两阶段提交，示例如下。

```
create subscription mysub
 connection 'hostaddr=127.0.0.1 port=1501 dbname=logical_src'
 publication mypub1 with(create_slot='false',slot_name='logical_slot1',two_phase =
    on);
```

注意，two_phase 选项只能在创建订阅时设置，不能修改。

2．允许发布模式下的所有表

在 PostgreSQL 15 中创建发布语法时，添加了 for tables in schema 选项，允许指定一个或多个模式，发布模式下的所有表，示例如下。

```
create publication mypub for tables in schema s1,s2;
```

除了允许指定模式，还允许同时发布一个或多个表，示例如下。

```
create publication mypub for table public.tab1, public.tab2, tables in schema
    s1,s2;
```

3．支持行级和列级过滤

在 PostgreSQL 15 中创建发布时，可以使用 where 子句过滤行，示例如下。

```
create publication mypub for table tab1 where (id > 100 and id <= 1000);
```

也可以只过滤部分列，示例如下。

```
create publication mypub for table tab1(a,b);
```

还可以对行和列进行组合发布，示例如下。

```
create publication uno for table t(a, b) where (a > 0);
create publication dos for table t(a, c) where (a < 0);
```

4．改进复制冲突处理

订阅端会因主键冲突等而引起复制冲突，在默认情况下，PostgreSQL 会在出现错误时不断重试。在 PostgreSQL 15 之前的版本中，有两种方式来处理复制冲突：一种是手动删除冲突数据来让复制继续，另一种是使用 pg_replication_origin_advance 函数将 LSN 推进到失败事务之后的位置，从而跳过冲突的事务，重启复制。

这两种方式都不太方便，尤其是第二种方式需要使用 pg_waldump 工具或发布端的其他工具找到失败事务的 LSN。

PostgreSQL 15 允许订阅端在出错时禁用订阅，示例如下。

```
alter subscription mysub set (disable_on_error = on);
```

使用 disable_on_error 选项，订阅端在出错时可以通过自动禁用订阅来打破循环，并且通过 pg_stat_subscription_stats 视图还可以观察到累计报错的次数。

此外，还可以使用一种更健壮的方式来跳过冲突的事务，示例如下。

```
alter subscription mysub skip (lsn = '0/8777BA80');
```

系统通过对用户指定的 LSN 执行检查来避免设置错误的 LSN。

6.5 PostgreSQL 15 的系统层变化

PostgreSQL 15 的系统层变化主要体现在系统元数据引入、系统函数变化、预置角色变化、配置参数变化、GRANT 命令授权变化、递归查询优化、公共模式安全性提高、视图安全性提高、附加模块变化。

6.5.1 系统元数据引入

PostgreSQL 15 新引入了 5 个系统元数据（2 个系统表和 3 个系统视图），如表 6-1 和表 6-2 所示。

表 6-1 PostgreSQL 15 新引入的系统表

系统表名称	描述
pg_parameter_acl	记录用户被授权的参数信息。PostgreSQL 15 可以对需要超级用户权限设置的参数通过 GRANT SET 命令和 GRANT ALTER SYSTEM 命令下放权限给普通用户，用户被授权的参数信息可以通过 pg_parameter_acl 表查询，新增的权限简称为 s 或 A
pg_publication_namespace	记录模式和发布之间的映射关系。PostgreSQL 15 中逻辑复制的发布语法现在支持对 Schema 模式下的所有表进行发布，模式和发布之间是多对多的关系，一个 Publication 可以发布多个 Schema，一个 Schema 也可被创建到多个 Publication 下

表 6-2 PostgreSQL 15 新引入的系统视图

系统视图名称	描述
pg_ident_file_mappings	记录 ident 认证方式的映射内容。pg_hba.conf 文件的配置内容可以通过 pg_hba_file_rules 视图查看，但是 pg_hba.conf 文件嵌套引用包含的 pg_ident.conf 文件不可以通过 pg_hba_file_rules 视图查看。PostgreSQL 15 增加了 pg_ident_file_mappings 视图，用于查看 pg_ident.conf 文件的详细映射内容。pg_ident_file_mappings 视图是通过 pg_ident_file_mappings 函数实现的
pg_stat_subscription_stats	记录逻辑复制订阅统计信息。pg_stat_subscription_stats 视图对订阅在初始同步表数据和应用时发生错误的次数进行统计。pg_stat_get_subscription_stats 视图是通过 pg_stat_get_subscription_stats 函数实现的

续表

系统视图名称	描述
pg_stat_recovery_prefetch	记录数据库恢复预取信息。PostgreSQL 15 新引入了参数 recovery_prefetch，该参数在数据库恢复期间，对 WAL 的处理增加了一个预读取的功能。它可以指导系统内核在恢复之前对 buffer pool 中没有的数据块发起磁盘数据块的读取功能，从而降低 I/O 等待时间。pg_stat_recovery_prefetch 视图是通过 pg_stat_get_recovery_prefetch 函数实现的

6.5.2 系统函数变化

PostgreSQL 15 中的系统函数变化主要如下。

1．新引入了 4 个支持正则表达式处理的函数

- regexp_count：统计字符串匹配某个子串的次数。
- regexp_instr：返回正则表达式第 N 次匹配的位置。
- regexp_like：检测字符串是否模糊匹配正则表达式。
- regexp_substr：返回正则表达式第 N 次匹配的子串。

2．新引入了 2 个系统信息函数

- pg_settings_get_flags：获取参数的内部标识。
- has_parameter_privilege：判断用户对 GUC 参数是否具有 SET 权限或 ALTER SYSTEM 权限。

3．新引入了 1 个监控统计函数

- pg_stat_reset_subscription_stats：重置 pg_stat_subscription_stats 视图的订阅统计信息。

4．新引入了 7 个系统管理函数

- pg_backup_start：基础备份开始函数，由 pg_start_backup 函数重命名而来。
- pg_backup_stop：基础备份结束函数，由 pg_stop_backup 函数重命名而来。
- pg_ls_logicalsnapdir：监控逻辑复制目录 pg_logical/snapshots 下的文件信息。
- pg_ls_logicalmapdir：监控逻辑复制目录 pg_logical/mappings 下的文件信息。
- pg_ls_replslotdir：监控复制槽目录 pg_replslot/slot_name 下的文件信息。
- pg_get_wal_resource_managers：查看 WAL 的资源管理器列表信息。
- pg_database_collation_actual_version：显示数据库运行时 Collation 的版本，检测 Collation 是否与操作系统对齐，从而避免数据损坏。

5．其他变化

- pg_size_bytes：增加对 PB 单位的支持。
- pg_size_pretty：增加对 PB 单位的支持。
- pg_stat_reset_single_table_counters：允许重置跨库共享系统表的统计信息。

- pg_log_backend_memory_contexts：允许使用 GRANT 命令赋予普通用户执行该函数的权限。

6.5.3 预置角色变化

PostgreSQL 近期的几个版本都在丰富数据库的预置角色，其核心理念是为了弱化对超级用户权限的依赖。在 PostgreSQL 15 之前的版本中，只有超级用户才能执行 CHECKPOINT 操作，普通用户在执行该操作时会有明确的错误提示。

```
postgres=# \c postgres dba
You are now connected to database "postgres" as user "dba".

postgres=> checkpoint;
ERROR: must be superuser to do CHECKPOINT
```

PostgreSQL 15 新增了 pg_checkpoint，此角色可以执行 CHECKPOINT 操作。

```
postgres=# grant pg_checkpoint to dba;
GRANT ROLE

postgres=# \c - dba
You are now connected to database "postgres" as user "dba".

postgres=> checkpoint;
CHECKPOINT
```

普通用户 dba 被赋予 pg_checkpoint 后，拥有了执行 CHECKPOINT 操作的权限，这个变化使得执行 CHECKPOINT 操作不再需要依赖超级用户权限，安全性得到了一定的提高。

PostgreSQL 15 支持的预置角色如表 6-3 所示。

表 6-3 PostgreSQL 15 支持的预置角色

预置角色名称	描述
pg_checkpoint	允许执行 CHECKPOINT 操作
pg_database_owner	提供数据库宿主的通用权限
pg_read_all_data	提供全局读取的访问权限
pg_write_all_data	提供全局写入的访问权限
pg_execute_server_program	可以执行数据库服务端程序以配合 COPY 操作和其他允许执行服务端程序的函数
pg_read_server_files	可以使用 COPY 操作及其他文件访问函数在数据库服务端可访问的任意位置读取文件
pg_write_server_files	可以使用 COPY 操作及其他文件访问函数在数据库服务端可访问的任意位置写入文件
pg_read_all_settings	可以读取所有配置变量，包括那些通常只对超级用户可见的信息
pg_read_all_stats	可以读取所有 pg_stat_* 视图信息，以及相关的扩展统计信息，包括那些通常只对超级用户可见的信息。PostgreSQL 15 新增了对 pg_backend_memory_context 视图和 pg_shmem_allocations 视图的访问

预置角色名称	描述
pg_stat_scan_tables	可以执行一些监控函数，需要获取表的 Access Share 锁
pg_monitor	可以执行各种监控视图和函数，是 pg_read_all_settings、pg_read_all_stats 和 pg_stat_scan_tables 的成员
pg_signal_backend	可以使用 pg_cancel_backend 函数或 pg_terminate_backend 函数对后端进程发送信号，取消后端进程查询或中止后端进程

6.5.4 配置参数变化

PostgreSQL 15 中配置参数的主要变化有：新引入了 8 个配置参数，修改了 5 个配置参数，并废弃了 1 个配置参数，如表 6-4～表 6-6 所示。

表 6-4　PostgreSQL 15 新引入的配置参数

参数名称	描述
archive_library	为归档操作添加了新的实现方式，可以通过加载 basic_archive 模块来代替参数 archive_command。相比参数 archive_command 使用 shell 命令来对每个文件进行归档，如果产生大量的 WAL，那么服务端需要等待本地 shell 命令的交互，参数 archive_library 的优势比较明显
log_startup_progress_interval	数据库启动恢复过程在一些场景中可能运行较慢，如系统需要花费一些时间应用 WAL，或花费较长时间重置 UNLOGGED 对象，或执行 fsync 同步数据到磁盘，但数据库日志并不能很好地体现在哪个阶段，设置参数 log_startup_progress_interval 后，当某个阶段耗时达到默认设置的 10s 后会更新该阶段的状态信息
recovery_prefetch	在数据库恢复期间，对 WAL 的处理增加一个预读取的功能，可以降低 I/O 等待时间，同时可以通过 pg_stat_recovery_prefetch 视图来观察
recursive_worktable_factor	允许用户在使用 WITH 语句进行递归查询的工作负载中设置 worktable 进行性能调整，可以预防因死循环而导致的问题
shared_memory_size	记录由参数 shared_preload_libraries 配置的扩展插件运行时的内存
shared_memory_size_in_huge_pages	显示参数 huge_page_size 的使用信息且与参数 shared_memory_size 相关的大页内存
stats_fetch_consistency	在 transaction 中多次访问累计的统计信息时，一致性读取可以有 3 个值：none、cache 、snapshot
wal_decode_buffer_size	如果打开了 WAL 预取参数，那么可以通过参数 wal_decode_buffer_size 设置 WAL 解码的缓存大小

表 6-5 PostgreSQL 15 修改的配置参数

参数名称	描述
hash_mem_multiplier	HASH 操作占用的内存超过 work_mem * hash_mem_multiplier 的值之后，剩余的内存占用会被溢出到磁盘。参数 hash_mem_multiplier 在 PostgreSQL 13 和 PostgreSQL 14 中的默认值为 1，在 PostgreSQL 15 中的默认值为 2，允许进行 HASH 操作。相比其他操作，进行 HASH 操作可以使用双倍的 work_mem
log_autovacuum_min_duration	默认值由-1变为10
log_checkpoints	默认值由 off 变为 on
log_destination	添加了 jsonlog 选项，支持 JSON 格式日志
wal_compression	在 PostgreSQL 15 之前的版本中，只能设置值为 off 和 on，当设置值为 on 时采用内置的 pglz 算法进行压缩，在 PostgreSQL 15 中除了可以设置值为 off 和 on，还可以设置值为 pglz、lz4 和 zstd

表 6-6 PostgreSQL 15 废弃的配置参数

参数名称	描述
stats_temp_directory	PostgreSQL 15 之前的版本通过统计信息收集进程更新统计信息，并将统计信息存储在文件系统参数 stats_temp_directory 配置的目录中。PostgreSQL 15 废弃了统计信息收集进程，将统计信息记录到共享内存中

此外，还有一个比较特殊的参数，即 allow_in_place_tablespaces。该参数在 PostgreSQL 15、PostgreSQL 14.5、PostgreSQL 13.8、PostgreSQL 12.12、PostgreSQL 11.17、PostgreSQL 10.22 中被引入。参数 allow_in_place_tablespaces 用于打开表空间功能，将数据直接存储到 pg_tblspc 目录下。该参数默认关闭，打开该参数后可以在同一台机器中非常方便地测试自定义表空间。

例如，当使用 pg_basebackup 工具进行备份时，可能会出现如下错误提示。

```
$ pg_basebackup --pgdata=databackup
pg_basebackup: error: directory "/…" exists but is not empty
pg_basebackup: removing data directory "databackup"
```

由于在同一台机器中复制 pg_tblspc 目录下的自定义表空间符号链接时会发生冲突，因此需要使用 --tablespace-mapping 选项来进行表空间目录映射。

当然，也可以通过设置参数 allow_in_place_tablespaces 进行处理。

```
postgres=# ALTER SYSTEM SET allow_in_place_tablespaces TO true;
ALTER SYSTEM

postgres=# SELECT pg_reload_conf();
 pg_reload_conf
----------------
 t
(1 row)
```

在创建表空间时，将参数 location 的值设置为空。

```
postgres=# create tablespace myspace location '';
CREATE TABLESPACE
```

此时，直接在 pg_tblspc 目录下生成表空间文件来存储数据。

PostgreSQL 中的参数有些是初始化时设置的，有些是从静态模板中继承的，还有些是运行时获取的。PostgreSQL 15 为用户提供了一个 pg_settings_get_flags 函数，用于获取参数的内部标识。

```
postgres=# select distinct flags
           from (select pg_settings_get_flags(name) as flags
                 from pg_settings) as x;
             flags
---------------------------------
 {EXPLAIN}
 {}
 {EXPLAIN,NOT_IN_SAMPLE}
 {NOT_IN_SAMPLE,RUNTIME_COMPUTED}
 {NOT_IN_SAMPLE}
 {NO_RESET_ALL,NOT_IN_SAMPLE}
(6 rows)
```

pg_settings_get_flags 函数返回的标识可以分为 5 种。

- EXPLAIN：EXPLAIN(SETTINGS)命令中包含的参数。
- NO_SHOW_ALL：SHOW ALL 命令中排除的参数。
- NO_RESET_ALL：RESET ALL 命令中排除的参数。
- NOT_IN_SAMPLE：不包含在 postgresql.conf 文件中的参数。
- RUNTIME_COMPUTED：运行时计算的参数。

要想查看哪些参数不会被 RESET ALL 命令重置，可以使用如下代码。

```
postgres=# with flags_all as
           (select name,pg_settings_get_flags(name) as flags from pg_settings)
           select name,flags
             from flags_all
            where flags_all.flags @> ARRAY['NO_RESET_ALL'];
         name           |            flags
------------------------+------------------------------
 transaction_deferrable | {NO_RESET_ALL,NOT_IN_SAMPLE}
 transaction_isolation  | {NO_RESET_ALL,NOT_IN_SAMPLE}
 transaction_read_only  | {NO_RESET_ALL,NOT_IN_SAMPLE}
(3 rows)
```

一个参数可能会有多个标识，用户可以根据 pg_settings_get_flags 函数获取的标识数组参数直接查询。要查看哪些参数是运行时计算的参数，可以使用多个标识数组参数。

```
postgres=# select name,setting
             from pg_settings
            where pg_settings_get_flags(name) ='{NOT_IN_SAMPLE,RUNTIME_COMPUTED}';
             name                 | setting
----------------------------------+---------
 data_checksums                   | off
 data_directory_mode              | 0700
 shared_memory_size               | 143
 shared_memory_size_in_huge_pages | 72
```

```
 wal_segment_size            | 16777216
(5 rows)
```

注意，在使用 x.y 形式的自定义参数时，参数分组的名称不能与数据库中已经安装的扩展插件的名称相同，否则扩展插件加载运行后会提示该参数的告警信息，系统会自动删除该参数。

6.5.5 GRANT 命令授权变化

在 PostgreSQL 15 之前的版本中，使用 SET 命令只能修改部分参数的值，有些需要超级用户权限设置的参数不能被修改，ALTER SYSTEM 命令也是如此。而在 PostgreSQL 15 中需要超级用户权限设置的参数可以通过 GRANT SET 命令和 GRANT ALTER SYSTEM 命令下放权限给普通用户，示例如下。

```
postgres=# GRANT SET on parameter log_checkpoints TO dba;
GRANT

postgres=# GRANT ALTER SYSTEM on parameter archive_mode TO dba;
GRANT
```

通过上面的设置可知，普通用户 dba 可以使用 SET 命令设置参数 log_checkpoints，也可以使用 ALTER SYSTEM 命令设置参数 archive_mode。

同时，被授权的参数权限可以通过 pg_parameter_acl 表查询，该表会记录所有被授权的参数权限。

```
postgres=# select parname,paracl from pg_parameter_acl ;
    parname      |              paracl
-----------------+------------------------------------
 log_checkpoints | {postgres=sA/postgres,dba=s/postgres}
 archive_mode    | {postgres=sA/postgres,dba=A/postgres}
(2 rows)
```

在上述代码的 sA、s、A 中，s 是 SET 权限的简称，A 是 ALTER SYSTEM 权限的简称。

has_parameter_privilege 函数也支持 SET 权限和 ALTER SYSTEM 权限，示例如下。

```
postgres=# select has_parameter_privilege('dba','log_checkpoints','set');
 has_parameter_privilege
-------------------------
 t
(1 row)

postgres=# select has_parameter_privilege('dba','archive_mode','alter system');
 has_parameter_privilege
-------------------------
 t
(1 row)
```

6.5.6 递归查询优化

使用 PostgreSQL 15 之前的版本进行递归查询的 work table 的平均记录数被评估为非

递归子句记录数的 10 倍，这个倍数固化在内核代码中。在不同的递归场景下，work table 的记录数的实际差异较大，对于递归的图式全部展开查询，每个 work table 的记录数可能是上一级的 N 倍。例如，如果每个人关联 10 个人，那么第一层是 10 条记录，第二层可能是 100 条记录。

PostgreSQL 15 新增了参数 recursive_worktable_factor，默认值依旧是 10。该参数允许在不同的递归场景下，通过查询计划器提供更好的计划来调整递归查询的工作负载性能。

在 PostgreSQL 15 中进行递归查询的示例如下。

```
with recursive t(n) as(
  values(300)
  union all
  select n+2 from t where n<500
) select avg(n) from t;
```

注意，如果递归语句与非递归语句混合使用，那么需要先编写非递归语句，再编写递归语句。

```
with recursive
  cte1 as (...)         -- cte1 为非递归语句
, cte2 as (select ...
           union all
           select ...)   -- cte2 为递归语句
, cte3 as (...)         -- cte3 为递归语句
select ... from cte3 where ...
```

6.5.7 公共模式安全性提高

长期以来，多数开发者都非常喜欢 PostgreSQL 中简单、易用的 public 模式。因为 PostgreSQL 默认允许用户连接任何数据库，并且在 public 模式下创建表、视图等对象，所以很多用户习惯把数据库对象放到 public 模式下。

在 PostgreSQL 14 中新建数据库后，可以看到用户默认拥有 public 模式的使用和创建权限。

```
postgres=# create database private_db;
CREATE DATABASE

postgres=# \c private_db
You are now connected to database "private_db" as user "postgres".

private_db=# \dn+ public
                  List of schemas
 Name  |  Owner   |    Access privileges    |     Description
--------+----------+-------------------------+------------------------
 public | postgres | postgres=UC/postgres+   | standard public schema
        |          | =UC/postgres            |
(1 row)
```

任何用户都可以在 public 模式下创建数据库对象。

```
private_db=# create user anyuser;
CREATE ROLE
```

```
private_db=# \c - anyuser
You are now connected to database "private_db" as user "anyuser".

private_db=> create table test(id int);
CREATE TABLE
```

有些用户认为这样使用 public 模式不太安全，会选择把 public 模式删除，同时包括删除通用的 public 角色拥有的 connect 权限，以提高安全性。

```
postgres=# revoke connect on database private_db from public;
REVOKE

postgres=# \c private_db anyuser
connection to server on socket "/tmp/.s.PGSQL.1405" failed: FATAL:  permission
    denied for database "private_db"
DETAIL:  User does not have CONNECT privilege.
Previous connection kept
```

在上面的代码中把 private_db 数据库的 connect 权限从 public 角色中删除了，如果不先删除 public 角色的 connect 权限，而直接删除 anyuser 角色的 connect 权限，那么不会生效。

```
postgres=# revoke connect on database private_db from anyuser;
REVOKE

postgres=# \c private_db anyuser
You are now connected to database "private_db" as user "anyuser".

private_db=>
```

下面观察 PostgreSQL 15 中的 public 模式的权限变化。

```
postgres=# create database otherdb;
CREATE DATABASE

postgres=# create database private_db with owner dba;
CREATE DATABASE

postgres=# \c private_db
You are now connected to database "private_db" as user "postgres".

private_db=# \dn+
                              List of schemas
 Name   |       Owner        |            Access privileges             |
  Description
--------+--------------------+------------------------------------------+-------------
  ------------
 public | pg_database_owner  | pg_database_owner=UC/pg_database_owner+  | standard
    public schema
        |                    | =U/pg_database_owner                     |
(1 row)
```

可以看出，PostgreSQL 15 中的 public 模式权限发生了两个变化，一是创建权限被删除了，二是 owner 被修改为 pg_database_owner。普通用户默认不能在 public 模式下创建对象，只有数据库的宿主才能创建。

```
$ psql -U dba -d otherdb -c 'create table test (id int)'
ERROR:  permission denied for schema public
LINE 1: create table test (id int)
                     ^

$ psql -U dba -d private_db -c 'create table test (id int)'
CREATE TABLE
```

otherdb 数据库的宿主是 postgres，private_db 数据库的宿主是 dba，用户 dba 只能在 private_db 数据库的 public 模式下成功创建表。

PostgreSQL 15 对 public 模式默认权限的修改是一个较大的变化，如果希望其能够保持与 PostgreSQL 15 之前的版本兼容，那么需要明确为 public 模式赋予创建权限。在创建数据库时，如果希望保持 PostgreSQL 15 之前的版本的兼容性，那么可以通过修改 template1 数据库来实现。

6.5.8 视图安全性提高

PostgreSQL 对函数的执行有两种安全上下文环境：一种是 security invoker，以调用者权限执行函数，这是默认的安全环境；另一种是 security definer，以创建者权限执行函数，创建者权限在函数执行期间生效。使用 security invoker，在普通用户定义的函数体中包含高危操作时，虽然普通用户没有权限执行，但如果被权限较高的用户执行，会触发函数陷阱。而 security definer 则可以把相关权限较高的操作封装限定，让用户在可控范围内执行。

PostgreSQL 的视图使用方法类似于 security definer，也就是说，PostgreSQL 使用视图的宿主检测相关对象的权限。

下面通过示例进行视图测试。

```
postgres=# create user test;
CREATE ROLE

postgres=# create table tab_base as select 1 as id;
SELECT 1

postgres=# create view my_view as select * from tab_base;
CREATE VIEW

postgres=# grant select on my_view to test;
GRANT

postgres=# \c - test

postgres=> select * from my_view;
 id
----
  1
(1 row)
```

在上面的示例中先创建了 tab_base 表，再创建了 my_view 视图，只需要把 my_view

视图的查询语句赋予测试用户，测试用户就可以访问到数据。使用视图的这个特性，可以减少视图中相关对象的权限设置，允许用户有部分访问权限，应注意 security_barrier 属性对视图攻击的安全保护，这一点在官方文档中有详细的示例说明。

PostgreSQL 15 对视图新增了一个 security_invoker 属性，默认处于关闭状态。下面修改创建的 my_view 视图，设置 security_invoker 属性的值为 on。

```
postgres=# alter view my_view set (security_invoker = on);
ALTER VIEW

postgres=# \c - test
You are now connected to database "postgres" as user "test".

postgres=> select * from my_view ;
ERROR:  permission denied for table tab_base
```

当设置 security_invoker 属性的值为 on 之后，系统将检查用户对视图相关对象（示例中的 tab_base 表）是否有权限，而不是只检查视图的 owner 权限。由于没有对 tab_base 表赋予权限给测试用户，因此查询语句也明确提示对 tab_base 表无权限。

视图引用的基表如果开启了行级安全特性，不打开 security_invoker 属性，那么行级安全策略会被跳过。因此，设置 security_invoker 属性后，视图能与行级安全特性匹配得更好。

6.5.9 附加模块变化

PostgreSQL 15 中附加模块的变化具体如下。
- 新增 basebackup_to_shell 模块，可以配合 pg_basebackup 工具的 target 选项使用。
- 新增 basic_archive 模块，用于新的归档实现。
- 如果参数 archive_library 的值为空，那么使用传统参数 archive_command 的 shell 命令进行归档，如果参数 archive_library 的值为 basic_archive，那么使用 basic_archive 模块进行归档，由于 basic_archive 模块底层通过 C 实现，因此使用 basic_archive 模块比使用 shell 命令的性能会好一些。如果同时设置了参数 archive_library 和 archive_command，那么参数 archive_command 的设置不会生效。
- 新增 pg_walinspect 模块。其功能类似于 pg_waldump 工具，可以通过 SQL 语句进行查看。
- postgres_fdw 模块在创建服务器时新增 parallel_commit 选项，支持并行提交，支持 case 表达式下推。
- pg_stat_statements 模块增加临时文件块的 I/O 耗时信息。

6.6 本章小结

PostgreSQL 15 在性能提升方面主要包括统计信息使用共享内存的方式直接集成到内

核中；增量排序算法改进；在 WAL 恢复期间通过预读取降低 I/O 等待延迟；全页写新增效率更高的 lz4 算法和 zstd 算法；备份支持服务端压缩及更细粒度的压缩控制选项，包括并行处理等。

PostgreSQL 15 在可靠性提高方面主要包括一致性读取事务中的统计信息；废弃不可靠的独占备份模式，引入统一非独占备份模式；本地化 Collation 可对齐刷新至操作系统等。

PostgreSQL 15 在运维管理优化方面主要包括引入服务端本地备份或通过插件扩展至云端存储；引入 JSON 格式日志，便于转储和分析；创建数据库功能增强；COPY 操作对文本格式增强；PSQL 工具优化等。

PostgreSQL 15 在开发易用性提升方面主要包括引入 Oracle 兼容的 MERGE 语句；允许开发人员控制 NULL 值是否重复；改进 numeric 数据类型，兼容 Oracle 的精度位与小数位设置；引入正则表达式函数；改进分区表等。PostgreSQL 15 在系统层变化方面主要包括递归查询优化；公共模式安全性提高；视图安全性提高；附加模块变化等。

第 7 章

PostgreSQL 16 新特性

7.1 PostgreSQL 16 的主要性能提升

与 PostgreSQL 16 之前的版本相比，PostgreSQL 16 的性能有了一定的提升，主要体现在并行特性增强、预排序性能提升、死元组清理性能提升、其他性能提升。

7.1.1 并行特性增强

PostgreSQL 16 引入了新的查询并行执行计划，可以更好地支持并行查询。并行查询的原理是把查询操作分成多个任务，将每个任务分配给不同的工作进程并行执行。在生成并行执行计划之前，优化器会考虑查询的复杂性、数据分布、数据量、需使用的系统资源等因素，并对具体各部分的执行流程进行资源分配。

PostgreSQL 16 的并行特性有如下方面的增强。

- 引入并行 HASH FULL JOIN 操作和 HASH RIGHT JOIN 操作。

PostgreSQL 11 新增了参数 enable_parallel_hash，支持 HASH JOIN 操作使用并行，但不支持 FULL JOIN 操作和 RIGHT JOIN 操作使用并行，PostgreSQL 16 支持这两种操作使用并行。

- string_agg 函数和 array_agg 函数支持并行聚合。

PostgreSQL 16 对 string_agg 函数和 array_agg 函数增加了内置的 serialize 函数、deserializ 函数、transfn 函数、combine 函数、finalfn 函数来支持并行聚合。当数据量比较大时，多个工作进程可以同时对不同数据页的子集执行聚合操作，并把各部分的结果合并，以加快整体聚合的速度。

- postgres_fdw 模块支持异步中止事务。

PostgreSQL 15 中的 postgres_fdw 模块新增了 parallel_commit 选项，对外部表提供了异步提交事务的能力，而 PostgreSQL 16 中的 postgres_fdw 模块新增了 parallel_abort 选项，对外部表提供了异步中止事务的能力。

7.1.2 预排序性能提升

PostgreSQL 16 中新增了参数 enable_presorted_aggregate。该参数允许优化器对带 ORDER BY 子句或 DISTINCT 子句的聚合操作通过预排序来减少显式排序的代价消耗。

下面是两种聚合场景的关键代码。

```
array_agg(a order by a), string_agg(a order by a,b)
string_agg(distinct a), array_agg(distinct a,b)
```

在 PostgreSQL 16 之前的版本中，每个聚合函数都要单独排序，数据量比较大时的排序性能可能会较差。在 PostgreSQL 16 中，优化器可以使用索引及预排序来优化 ORDER BY 子句或 DISTINCT 子句。不过，在数据分组分布不均匀的情况下，使用预排序执行 ORDER BY 子句或 DISTINCT 子句，速度反而会变慢，此时可以禁用参数 enable_presorted_aggregate，使用 PostgreSQL 16 之前的版本中的默认行。

7.1.3 死元组清理性能提升

在 PostgreSQL 16 中，死元组清理性能有如下方面的提升。

- AUTOVACUUM 进程刷新代价相关的参数更积极。

AUTOVACUUM 进程相比 VACUUM 操作有一定的限速限流，这也是为了降低对系统 I/O 的影响。当大表的自动清理操作处理时间比较长时，可以通过修改代价相关的参数 autovacuum_vacuum_cost_limit 和 autovacuum_vacuum_cost_delay 加快处理速度。

在 PostgreSQL 16 之前的版本中，自动清理进程只有等当前表处理完成之后才会加载并更新参数值。在 PostgreSQL 16 中，受参数 autovacuum_vacuum_cost_delay 的控制，只要下一个周期检测到参数值变化就会加载并更新参数值。

- VACUUM 操作使用的 ring buffer 内存可控。

shared buffer 内存中用于 VACUUM 操作或 ANALYZE 操作的 ring bufer 内存在 PostgreSQL 16 之前的版本中是固定的 256KB，当对大表执行 VACUUM 操作时，后台处理速度非常缓慢。

PostgreSQL 16 新增了参数 vacuum_buffer_usage_limit。同时，在 VACUUM 命令的语法中也增加了 buffer_usage_limit 选项，在针对大表执行 VACUUM 操作时可以配置大一些的 ring bufer 内存来加快处理速度。

7.1.4 其他性能提升

PostgreSQL 16 还有一些其他性能得以提升，具体如下。

- 数据文件扩展操作优化，并发写入性能提升。

当大量记录频繁添加到表中时，表对象的物理数据文件不得不追加新页进行扩展操作。为了预防多个客户端同时添加新页，客户端只有获取对象上的扩展锁后才能进行操作，而扩展锁的使用会显著影响多个活跃客户端添加记录的性能。

PostgreSQL 16 通过 13 处独立的修改，降低了持有对象扩展锁的层级，这对多个客户端频繁插入记录会有较大的好处，同时单个客户端的插入记录的花销也会降低。

- 并发 COPY 操作的性能提升。

受数据文件扩展操作的影响，PostgreSQL 16 在并发 COPY 操作的性能上也有了一定的提升。

- 配置参数存储及访问优化。

PostgreSQL 16 优化了配置参数存储并加快了对参数的访问速度，当在函数中或存储过程中大量使用自定义参数时，参数的访问速度会有较大的提高。

以下代码配置了一万个自定义参数，在 PostgreSQL 16 之前的版本中首次执行时大约需要 4~6s，而在 PostgreSQL 16 中执行时大约需要几十毫秒。

```
do $$
    begin
        for i in 1..10000 loop
        perform set_config('foo.bar' || i::text, i::text, false);
        end loop;
end $$;
```

- 索引优化。

在构建 HASH 索引时，PostgreSQL 16 不仅按照 HASH 桶的编号进行排序，而且按照 HASH 索引值的顺序存储索引页，这样既可以避免索引页中的重复数据移动，又可以缩短 HASH 索引初始化构建的时间。

7.2 PostgreSQL 16 的运维管理优化

PostgreSQL 16 的运维管理优化主要体现在 I/O 统计更详细、pg_dump 工具功能增强、PSQL 工具功能增强、VACUUM 命令及 VACUUMDB 工具功能增强、pg_hba.conf 文件配置更高效、普通用户连接优化、HOT 更新监控增强、不活跃索引监控，以及便捷的参数化语句分析。

7.2.1 I/O 统计更详细

在 PostgreSQL 16 之前的版本中有如下视图可以查看 I/O 信息。
- pg_stat_database：

hits,reads,read time,write time。
- pg_statio_all_tables(pg_statio_sys_tables/pg_statio_user_tables)：

hits,reads。
- pg_stat_bgwriter：

backend writes,backend fsyncs。
- pg_stat_statements：

shared and local buffer hits,reads,writes,read time,write time。

使用以上视图可以分别从数据库、后台写进程、语句 3 个维度粗略地统计 I/O 信息、缓存命中、刷盘及时间消耗。不过，I/O 信息比较分散，并且 I/O 信息依赖于系统层的数据。

PostgreSQL 16 引入了一个非常重要的统计视图，即 pg_stat_io 视图。pg_stat_io 视图对 I/O 的统计粒度比较细，分为 3 个大的标签维度，即进程类型（backend_type 字段）、关系类型（object 字段）、操作上下文类型（context 字段），以及如下 8 个小的指标维度。

- reads/read_time：读操作及花费的时间。
- writes/write_time：写操作及花费的时间。
- writebacks/writeback_time：请求操作系统内核持久化回写存储的次数及花费的时间。
- extends/extend_time：因表数据文件空间不足而扩展页的次数及花费的时间。
- hits：命中 shared buffer 数据块的次数。
- evictions：shared buffer 或 local buffer 因未找到可保留的数据块而必须淘汰置换一个数据块的次数。
- reuses：在 BULKREAD 操作、BULKWRITE 操作、VACUUM 操作类型上重用环形缓冲区的次数。
- fsyncs/fsync_time：fsync 系统调用的次数及花费的时间（只统计 normal context）。

基于上面 3 个大的标签维度及 8 个小的指标维度，可以通过数据库清晰地观测到 I/O 信息。例如，观测到写压力是由数据大量写入而对表数据文件扩展页引起的，这是正常行为。如果是由刷盘写入引起的，那么可以对参数进行调优。

7.2.2　pg_dump 工具功能增强

PostgreSQL 16 对 pg_dump 工具的功能进行了如下增强。

- --compress 选项新增 lz4 算法和 zstd 算法。

与 pg_basebackup 工具用法类似，--compress 选项可以指定压缩方法。

```
$ pg_dump -f tmp.sql --compress=lz4
$ pg_dump -f tmp.sql --compress=lz4:level=5
```

- 新增--table-and-childrenget 选项，可以单独导出分区表结构。

在 PostgreSQL 16 之前的版本中，导出分区表结构有两种方式：一种方式是配置多个参数-t 显式指定所有分区，另一种方式是使用百分号通配符。上面这两种方式并不完美：使用第一种方式，表分区的数量可能并不是固定不变的；使用第二种方式，模糊匹配名可能与其他表名重叠。PostgreSQL 16 新增的--table-and-children 选项可以直接导出所有分区，包括声明式分区和继承表，代码如下。

```
$ pg_dump -f tmp.sql -s --table-and-children=tab
```

同时，PostgreSQL 16 新增了 --exclude-table-and-children 选项和 --exclude-table-data-and-children 选项，可以排除分区表的结构或数据。
- 表级共享访问锁批量发送，减少加锁交互时长。

7.2.3　PSQL 工具功能增强

PostgreSQL 16 对 PSQL 工具的功能进行了如下增强。
- 新增元命令\bind，支持扩展查询协议。
- 新增元命令\drg，支持查看成员角色 grant 信息。
- 元命令\watch 新增参数用于控制命令重复执行的次数。

在 PostgreSQL 16 之前的版本中，元命令\watch 只能使用 interval 选项控制命令间隔执行的秒数，代码如下。

```
\watch interval=3
```

PostgreSQL 16 新增 count 选项，可以控制命令重复执行的次数，代码如下。

```
\watch interval=3 count=2
```

interval 和 count 也可以简写为 i 和 c，代码如下。

```
\watch i=3 c=2
```

- 新增变量用于显示操作系统命令执行的情况。

新增变量 SHELL_ERROR 及 SHELL_EXIT_CODE。

例如，下面的代码中使用的是元命令\!执行操作系统命令 pwd 查看当前路径。

```
postgres=# \! pwd
/home/postgres
postgres=# \echo :SHELL_ERROR
false
postgres=# \echo :SHELL_EXIT_CODE
0
```

变量 SHELL_ERROR 的值为 false，表示执行操作系统命令 pwd 没有错误。
变量 SHELL_EXIT_CODE 的值为 0，表示执行操作系统命令 pwd 正常返回 0。

7.2.4　VACUUM 命令及 VACUUMDB 工具功能增强

PostgreSQL 16 对 VACUUM 命令及 VACUUMDB 工具的功能进行了如下增强。
- VACUUM 命令（包括 analyze 命令）新增 buffer_usage_limit 选项用于控制 buffer cache 的使用。

针对大表的 VACUUM 操作可以通过设置 buffer_usage_limit 选项控制 buffer cache 的使用，当 buffer_usage_limit 选项的值为 0 时表示允许使用所有 shared buffer，代码如下。

```
vacuum(analyze, buffer_usage_limit 0) tab;
```

- 使用 VACUUM 命令可以单独刷新数据库统计信息。

在 PostgreSQL 16 之前的版本中如果 VACUUM 操作的表很多，那么每次都会对表刷新数据库维度的 FrozenXID（冻结事务号）。在 PostgreSQL 16 中可以使用 skip_database_stats 选项先不刷新数据库维度的 FrozenXID，而使用 only_database_stats 选项只进行一次刷新操作。
- 使用 VACUUM 命令及 VACUUMDB 工具可以快速清理 TOAST 表。

VACUUM 命令可以使用 process_main 选项只处理主表数据或跳过主表数据，也可以使用 process_toast 选项只处理附属的 TOAST 表数据或跳过 TOAST 表数据，这也是通过主表来清理 TOAST 表的一种快捷方式，否则需要从 pg_class 表中进行关联查询以获取 TOAST 表的名称。

VACUUMDB 工具可以使用 --no-process-main 选项跳过处理主表数据，或使用 --no-process-toast 选项跳过处理 TOAST 表数据。

- VACUUMDB 工具可以指定模式或排除模式。

VACUUMDB 工具可以使用 --schema 选项只处理某些模式下的对象，也可以使用 --exclude-schema 选项跳过处理某些模式下的对象，以便进行模式为多租户场景下的清理工作。

7.2.5 pg_hba.conf 文件配置更高效

在 PostgreSQL 16 之前的版本中，pg_hba.conf 文件的 database 项和 user 项都可以配置一个或多个值（多个值之间用逗号分隔）。下面配置 database 项的代码。

```
local   mydb1,mydb2,postgres   all   trust
```

而配置 user 项，在 PostgreSQL 16 之前的版本中也有较多的方式。
- 配置一个用户或用逗号分隔的多个用户。
- 使用"+"配置角色或成员前缀。
- 使用"@"配置用户列表文件路径。

配置 user 项的代码如下：

```
local   mydb1   postgres,admin,+dbgrp,@/home/postgres/users.lst   trust
```

在 PostgreSQL 16 中可以使用正则表达式更加简洁地配置 database 项和 user 项。

```
local   "/^mydb[0-9]+$,/"   all   trust
host    all   "/_readonly$"   all   scram-sha-256
```

第一行代码使用正则表达式匹配以 mydb 为前缀并带数字的 database，第二行代码使用正则表达式匹配以 _readonly 结尾的 user。

PostgreSQL 16 还支持在 pg_hba.conf 文件中使用关键字 include，可以使用以下 3 种方式包含文件或目录。

- include pg_hba_extra.conf。
- include_if_exists pg_hba_extra.conf。
- include_dir hba_conf。

前两种方式允许包含文件，第 3 种方式允许包含目录。

7.2.6 普通用户连接优化

在 PostgreSQL 16 之前的版本中，数据库中最大的连接数由参数 max_connections 控制，该参数控制的连接数既包含参数 superuser_reserved_connections 控制的预留给超级用户使用的连接数,又包含参数 max_wal_senders 控制的流复制的连接数。

这两部分可以独立配置的连接数用于保证数据库之间的流复制、基础备份,以及一些后台运维操作不会因高并发情况下应用程序连接膨胀耗尽而无法正常工作。

数据库的一些后台运维操作虽然可以使用预留给超级用户使用的连接数,但是这并不是一个妥当的方式。例如,流复制连接也可以使用普通用户权限进行操作,当主库连接数占满时,备库无法和主库保持同步。另外,越来越多的运维和监控操作已不再依赖超级用户。

PostgreSQL 16 新增了一个参数 reserved_connections,普通用户可以作为 pg_use_reserved_connections 的成员通过使用该参数预留的连接数来保证日常运维和监控操作不会因应用程序连接膨胀耗尽而无法正常工作。

另外,数据库用户在 PostgreSQL 16 中获取连接的顺序分为如下几种情况。
- 普通用户:

只能使用除参数 reserved_connections 和 superuser_reserved_connections 外的连接数。
- pg_use_reserved_connections 的成员用户:

先使用普通用户的连接数,再使用参数 reserved_connections 预留的连接数,不能使用参数 superuser_reserved_connections 预留的连接数。
- 超级权限用户:

先使用普通用户的连接数,再使用参数 reserved_connections 预留的连接数,最后使用参数 superuser_reserved_connections 预留的连接数。

7.2.7 HOT 更新监控增强

在 PostgreSQL 的多个版本的实现中,数据项在堆表页中有多个版本而索引页中只有单个版本。当发生 UPDATE 操作并且更新非索引键的字段时,新的行版本直接追加到堆表的数据块中,而旧的行版本被标记为不可见。同时,也需要增加索引元组。当新的行版本与旧的行版本在同一个数据块中时,PostgreSQL 不会增加相应的索引元组,旧的行版本被标记为 Heap Hot Updated,新的行版本被标记为 Hot Only Tuple(HOT),这就是 HOT 机制。

HOT 机制由新的行版本与旧的行版本组成链表,用索引元组指向旧的行版本,并通过旧的行版本指向新的行版本,HOT 机制对多个版本的写的放大问题有较大程度的优化。

在 PostgreSQL 16 中,pg_stat_*_tables 视图家族新增了 n_tup_newpage_upd 字段,通过该字段可以对表进行 HOT 调优。例如,受表的 fillfactor 数据块填充率的影响,数据块预留空间不够,需要写入新的数据块,同时索引需要生成新的索引元组。此时,会发现 n_tup_newpage_upd 字段的值比较大,应该考虑降低表的填充率从而预留更多的空间用于 HOT 机制的更新。

7.2.8 不活跃索引监控

针对某个查询语句,查询规划器是否使用索引需要考虑如下因素:与大版本相关的内置算法、数据库配置参数、表的动态统计信息、使用索引的代价评估。不活跃索引是指查询语句很少或不被使用的索引。不活跃索引的使用对磁盘空间、数据膨胀及流复制都带来了很大的负担。不活跃索引产生的原因可能是大版本升级、配置文件调优、硬件升级、表重构或数据访问方式发生变化等。

在 PostgreSQL 16 之前的版本中，使用下面的语句可以查看不活跃索引。

```
select schemaname || '.' || indexrelname as index, idx_scan
 from  pg_stat_user_indexes;
```

在 PostgreSQL 16 中，pg_stat_*_tables 视图和 pg_stat_*_indexes 视图家族新增了两个字段：last_seq_scan 字段和 last_idx_scan 字段。使用 last_idx_scan 字段可以更加方便地查看最近哪些索引不活跃。

```
select schemaname || '.' || indexrelname as index, idx_scan, last_idx_scan
 from  pg_stat_user_indexes;
```

此外，还可以查看最近一周不活跃的索引。

```
select schemaname || '.' || indexrelname as index, idx_scan, last_idx_scan
from pg_stat_user_indexes
where last_idx_scan is null or last_idx_scan < now() - interval '7 days';
```

7.2.9　便捷的参数化语句分析

使用扩展查询协议可以分离 SELECT 命令、INSERT 命令、UPDATE 命令、DELETE 命令中使用的常量，进行参数化设置。使用参数化语句能够提高安全性，这是因为使用参数化语句可以防止 SQL 语句注入。同时，使用参数化语句也能提升系统性能，这是因为根据参数 plan_cache_mode 的默认值，数据库可以在单个会话中基于 6 次启发式训练来缓存参数化语句的通用执行计划。

在偏 OLAP 分析型的场景下运行耗时的查询语句时，获得最佳的执行计划变得尤为重要，此时可以把参数 plan_cache_mode 的值调整为 force_custom_plan，避免因执行计划缓存导致每次重新生成执行计划。

当在日常的数据库日志或 pg_stat_statements 视图中发现问题语句时，为了通过执行计划分析参数化语句的性能，可以设置参数 plan_cache_mode 的值为 force_generic_plan，借助通用执行计划来忽略变量的值。

下面使用 PREPARE 命令创建参数化语句。

```
PREPARE stmt(unknown) AS SELECT oid FROM pg_class WHERE relname = $1;
```

上面的语句使用伪类型 unknown，稍后可以让数据库根据上下文解析合适的数据类型。下面设置参数 plan_cache_mode 的值为 force_generic_plan。

```
SET plan_cache_mode = force_generic_plan;
```

使用如下命令查看执行计划。

```
EXPLAIN EXECUTE stmt(NULL);
```

在 PostgreSQL 16 中可以使用 EXPLAIN 命令的 generic_plan 选项，以更加简便的方式获取通用执行计划，完整的代码如下。

```
postgres=# EXPLAIN(generic_plan) SELECT oid FROM pg_class WHERE relname = $1;
                              QUERY PLAN
-------------------------------------------------------------------------
```

```
Index Scan using pg_class_relname_nsp_index on pg_class  (cost=0.27..8.29 rows=1
  width=4)
   Index Cond: (relname = $1)
(2 rows)
```

除此之外，数据库日志文件中的慢 SQL 语句还可以通过外部的日志分析工具 pgBadger，使用--dump-all-queries 选项提取参数的值并将其自动替换到参数化语句中进行分析。

7.3 PostgreSQL 16 的开发易用性提升

PostgreSQL 16 的开发易用性提升主要体现在逻辑复制功能完善、SQL 标准 2023 部分支持引入、SQL/JSON 函数功能增强、数据导入默认值重定义、libpq 协议负载均衡功能引入。

7.3.1 逻辑复制功能完善

在 PostgreSQL 14 及 PostgreSQL 15 中对逻辑复制的性能及功能有了大量的改进，如在 PostgreSQL 14 中对逻辑复制的大量性能进行了优化，包括支持流式处理大事务、提升 TRUNCATE 操作的性能、支持以二进制形式传输数据及优化表的初始数据同步等，在 PostgreSQL 15 中又对逻辑复制的功能进行了改进，包括支持两阶段提交、允许发布模式下的所有表等。

PostgreSQL 16 对逻辑复制的功能做了进一步完善。
- 支持 standby 节点逻辑解码。

PostgreSQL 16 可以通过新的 pg_log_standby_snapshot 函数捕获正在进行事务中的快照并将其存储到 WAL 文件中以支持备库逻辑解码，从而避免对检查点的依赖。
- 并行应用大事务。

PostgreSQL 16 通过新增的参数 max_parallel_apply_workers_per_subscription 控制订阅端并行应用大事务的工作进程数。
- 表初始数据同步以二进制形式复制行。

PostgreSQL 16 之前的版本虽然支持以二进制形式传输数据，但初始数据同步还是使用 COPY 的文本格式进行传输的，在 PostgreSQL 16 中创建订阅时通过 binary 选项可以加快初始数据同步速度，代码如下。

```
CREATE SUBSCRIPTION mysub WITH (binary);
```
- 订阅端允许以 owner 表执行数据操作。

PostgreSQL 16 的普通用户可以使用 pg_create_subscription 以 owner 表执行基本的操作来提升安全性，代码如下。

```
CREATE SUBSCRIPTION mysub WITH (run_as_owner);
```
- 复制标识允许使用非唯一索引。

PostgreSQL 16 允许逻辑复制在没有主键的情况下，使用非唯一索引用作复制标识进行应用。

- 解决循环复制。

PostgreSQL 16 可以区分数据变化是由用户的 SQL 语句引起的还是由 replication 引起的，在创建订阅时可以设置发布端不发布 replication 源，以避免陷入复制循环，代码如下。

```
CREATE SUBSCRIPTION mysub  PUBLICATION mypub1 WITH (origin = none) ;
```

- 实时逻辑解码。

PostgreSQL 16 新引入了参数 debug_logical_replication_streaming，用于控制逻辑解码日志的时机，该参数的值可以为 buffered 或 immediate，默认值为 buffered，保持 PostgreSQL 16 之前的版本的兼容性，逻辑解码需要先等待参数 logical_decoding_work_mem 的值填满后再解析。

开发测试阶段没必要等待产生足够的变化直到填满参数 logical_decoding_work_mem 的值进行解析，此时可以将参数 debug_logical_replication_streaming 的值设置为 immediate，以便进行实时解析。

7.3.2 SQL 标准 2023 部分支持引入

相比其他主流数据库甚至商业数据库，PostgreSQL 对 SQL 标准 2023 的支持更加全面，PostgreSQL 16 对 SQL 标准 2023 新增了如下支持。

- Non-decimal integer literals。

PostgreSQL 支持的字符串常量的形式非常丰富，支持的数字常量的形式比较少，PostgreSQL 16 支持十六进制形式、八进制形式、二进制形式的整型常量，使用非十进制形式更容易阅读和理解代码，如 4 字节整型最大值可以很方便地使用十六进制形式的 8 个 F 表示。相比十六进制形式，使用十进制形式更加不容易被识记。

下面分别使用十进制形式、十六进制形式、八进制形式和二进制形式输入 1234。

```
postgres=# SELECT 1234 dec_int,0x4D2 hex_int,0o2322 oct_int,0b10011010010 bin_int;
 dec_int | hex_int | oct_int | bin_int
---------+---------+---------+---------
    1234 |    1234 |    1234 |    1234
(1 row)
```

- Underscores in numeric literals。

PostgreSQL 16 可以使用下画线对整型常量和数字常量进行虚拟分组。

下面分别在测试函数、查询条件和更新设置时使用。

```
SELECT count(*) FROM generate_series(1, 1_000_000) ;
SELECT ... WHERE a > 0 and a < 3.14159_26535_89793/2 ;
UPDATE ... SET x = 0x_FFFF_FFFF ... ;
```

- ANY_VALUE。

PostgreSQL 16 实现的 any_value 函数可以对分组后的数据去重只返回单行。

下面代码中的 person 表，按部门编号各有两行数据。

```
postgres=# SELECT * FROM person;
 deptno |   name
```

```
--------+--------------
     10 | dept1_name1
     10 | dept1_name2
     20 |
     20 | dept2_name2
(4 rows)
```

查询 person 表，每个部门获取一位人员，使用 any_value 函数分组查询，部门编号为 10，返回第 1 行数据；部门编号为 20，跳过空行，返回第 4 行数据。

```
postgres=# SELECT deptno, any_value(name) FROM person GROUP BY deptno;
 deptno | any_value
--------+--------------
     10 | dept1_name1
     20 | dept2_name2
(2 rows)
```

在使用 any_value 函数时也可以指定排序。下面按姓名降序返回数据。

```
postgres=# SELECT deptno, any_value(name order by name desc)
        FROM person
        GROUP BY deptno;
 deptno | any_value
--------+--------------
     10 | dept1_name2
     20 | dept2_name2
(2 rows)
```

使用 any_value 函数可以在 select 列表中包含非聚合列而不会影响 GROUP BY 的正常行为，否则会出现如下错误提示。

```
ERROR: column "XXX" must appear in the GROUP BY clause or be used in an aggregate
  function
```

any_value 函数不仅提供了查询的便利性（不强制要求必须包含到 GROUP BY 中），而且从每个分组中只获取一个值，降低了计算成本，也提升了性能。

7.3.3　SQL/JSON 函数功能增强

PostgreSQL 15 提交了大量的 JSON 函数，最终正式发布前的回归测试检测因没有对 JSON 数据类型进行输入有效性检测而回退。

PostgreSQL 16 新引入了两个校验检测函数，即 pg_input_is_valid 函数和 pg_input_error_info 函数，第一个函数可以检测数据类型是否有效，如果输入值与数据类型不匹配，那么可以使用第二个函数查看错误信息。先检测数据类型再使用可以避免由数据类型转换失败导致事务中断的情况发生。

下面是被延迟发布到 PostgreSQL 16 中的 JSON 构造函数、聚合函数及断言函数。

- 构造函数 json_array。

对 JSON 对象构造 JSON 数组，可以使用 json_array 函数进行嵌套。

```
postgres=# SELECT json_array(100, json_array('Hello','PostgreSQL','16'),
  json('{}') );
             json_array
------------------------------------------
```

```
[100, ["Hello", "PostgreSQL", "16"], {}]
(1 row)
```

还可以在 json_array 函数中使用关键字 returning 返回 JSONB 数据类型。

```
postgres=# SELECT json_array(100, json_array('Hello','PostgreSQL','16'), json('{}')
    returning jsonb ) \gdesc
   Column    | Type
-------------+-------
 json_array  | jsonb
(1 row)
```

- 构造函数 json_object。

可以使用 JSON 的键值对构造 JSON 对象，同时结合使用 json_array 函数。

```
postgres=# SELECT json_object('ver': 16, 'bver':
    json_array('beta1','beta2','beta3'));
                   json_object
--------------------------------------------------
 {"ver" : 16, "bver" : ["beta1", "beta2", "beta3"]}
(1 row)
```

也可以在查询语句中使用 json_object 函数返回 JSONB 数据类型。

```
postgres=# SELECT json_object(usesysid: usename  returning JSONB) FROM pg_user
    LIMIT 1;
        json_object
------------------------
 {"16386": "u_readonly"}
(1 row)
```

- 聚合函数 json_arrayagg。

可以直接使用 json_arrayagg 函数返回 JSON 数组，不需要先使用子查询返回 json_array 函数。

```
postgres=# SELECT json_arrayagg(datname ORDER BY datname) FROM pg_database;
              json_arrayagg
------------------------------------------------
 ["mydb", "postgres", "template0", "template1"]
(1 row)
```

- 聚合函数 json_objectagg。

可以直接使用 json_objectagg 函数返回 JSON 对象。

```
postgres=# SELECT json_objectagg(usesysid: usename) FROM pg_user;
                      json_objectagg
------------------------------------------------------------------
 { "16386" : "u_readonly", "24686" : "admin", "10" : "postgres" }
(1 row)
```

- 断言函数 IS [NOT] JSON/JSON ARRAY/JSON OBJECT/JSON SCALAR。

可以使用 IS JSON 函数或 IS NOT JSON 函数判断文本值是否为 JSON 数据类型。

```
postgres=# SELECT '16' IS JSON, 'PostgreSQL' IS JSON;
 ?column? | ?column?
----------+----------
 t        | f
(1 row)
```

也可以使用 pg_input_is_valid 函数检测数据类型是否有效。

```
postgres=# select pg_input_is_valid('100','json');
 pg_input_is_valid
-------------------
 t
(1 row)
```

还可以结合 pg_input_error_info 函数查看错误信息。

```
postgres=# select pg_input_is_valid('PostgreSQL','json');
 pg_input_is_valid
-------------------
 f
(1 row)

postgres=# select pg_input_error_info('PostgreSQL','json');
                       pg_input_error_info
---------------------------------------------------------------------------
 ("invalid input syntax for type json","Token ""PostgreSQL"" is invalid.",,22P02)
(1 row)
```

这样就可以很方便地检测 JSON 数据类型了。

```
postgres=# SELECT f
            ,f IS json array AS json_array
            ,f IS json object AS json_object
            ,f IS json scalar AS json_scalar
        FROM unnest(array['16', '"PG"', '[3,6,9]', '{"ver":16}']) f;
     f      | json_array | json_object | json_scalar
------------+------------+-------------+-------------
 16         | f          | f           | t
 "PG"       | f          | f           | t
 [3,6,9]    | t          | f           | f
 {"ver":16} | f          | t           | f
(4 rows)
```

在 IS JSON 函数中还可以使用 WITH UNIQUE KEYS 或 WITHOUT UNIQUE KEYS 判断是否有重复的项。

```
postgres=# SELECT f
            ,f IS JSON OBJECT WITH UNIQUE KEYS AS obj_with
            ,f IS JSON OBJECT WITHOUT UNIQUE KEYS AS obj_without
            ,f IS JSON ARRAY WITH UNIQUE KEYS AS array_with
            ,f IS JSON ARRAY WITHOUT UNIQUE KEYS AS array_without
        FROM unnest(array['{"a":1, "b":2}',
                    '{"a":1, "a":2}',
                    '{"a":1, "b":{"c":2, "d":3}}',
                    '{"a":1, "b":{"c":2, "c":3}']) f;
              f              | obj_with | obj_without | array_with | array_without
-----------------------------+----------+-------------+------------+---------------
 {"a":1, "b":2}              | t        | t           | f          | f
 {"a":1, "a":2}              | f        | t           | f          | f
 {"a":1, "b":{"c":2, "d":3}} | t        | t           | f          | f
 {"a":1, "b":{"c":2, "c":3}} | f        | f           | f          | f
(4 rows)
```

7.3.4 数据导入默认值重定义

在使用 COPY FROM 命令导入数据文件时，假设数据文件中的某些数据项不想在导入时被设置为 NULL 而想使用表字段定义的默认值来代替。这种场景需求在 PostgreSQL 16 中可以使用 COPY FROM 命令的 default 选项实现。例如，在将时间字段导入到数据库中时将该字段的值设置为服务端的导入时间。

下面通过示例进行演示。

下面是 t1 表的结构代码。

```
CREATE TABLE t1(
a int,
b text default 'unknown',
c date default current_date
);
```

准备导入的数据文件 t1.csv 的内容如下（注意每行末尾不要有空格）。

```
1,value,2022-07-04
2,\D,2022-07-03
3,\D,\D
```

下面使用 COPY FROM 命令的 default 选项进行导入。

```
postgres=# \copy t1 from t1.csv with(format 'csv' ,default '\D')
COPY 3
```

当数据文件中包含 default 选项指定的字符串'\D'时，以表字段定义的默认值代替。

```
postgres=# select * from t1;
 a |   b     |     c
---+---------+------------
 1 | value   | 2022-07-04
 2 | unknown | 2022-07-03
 3 | unknown | 2023-07-16
(3 rows)
```

可以看出，t1 表中第二行数据的 b 字段的值，以及第三行数据的 b 字段和 c 字段的值都被替换为了默认值。

除了 COPY FROM 命令引入了 default 选项，外部表也引入了 default 选项，代码如下。

```
create extension file_fdw ;
create server file_server foreign data wrapper file_fdw;
create foreign table copy_default (
a integer,
b text default 'unknown',
c date default '2022-01-01'
) server file_server
options (format 'csv', filename '/home/postgres/t1.csv', default '\D');
```

上面的语句中在创建外部表的同时也引入了 default 选项。下面查看数据。

```
postgres=# select * from copy_default;
 a |   b   |     c
---+-------+------------
 1 | value | 2022-07-04
```

```
 2 | unknown | 2022-07-03
 3 | unknown | 2022-01-01
(3 rows)
```

与第一个示例类似，外部表 copy_default 的第二行数据的 b 字段的值，以及第三行数据的 b 字段和 c 字段的值都被替换为了默认值。

7.3.5　libpq 协议负载均衡功能引入

在 PostgreSQL 16 中，libpq 协议新引入两个参数。
- require_auth：允许客户端的 libpq 协议指定多种认证方式让服务端根据 pg_hba.conf 文件的配置来使用任意一种。新增的环境变量 PGREQUIREAUTH 有相同的作用。
- load_balance_hosts：控制客户端与多个 PostgreSQL 实例进行数据库连接的顺序。参数的默认值为 disable，保持 PostgreSQL 16 之前的版本的兼容性，另外一个值为 random，能以随机的方式连接任意一个 PostgreSQL 实例。

其中，使用参数 load_balance_hosts 对开发人员的帮助很大，使用 PostgreSQL 16 之前的版本也能与多个 PostgreSQL 实例进行连接，但不能控制实例节点的顺序，代码如下。

```
$ psql "host=node1,node2,node3"
```

第一个节点的压力比其他两个节点的压力大很多，而第三个节点甚至可能会一直处于空闲状态。

在 PostgreSQL 16 中可以使用新的参数 load_balance_hosts 进行连接。

```
$ /opt/pg16/bin/psql "host=node1,node2,node3 load_balance_hosts=random
  connect_timeout=15 target_session_attrs=standby "
```

在上面的示例中，参数 load_balance_hosts 的值为 random，可以对 host 列表的节点进行随机连接，结合 PostgreSQL 14 新引入的参数 target_session_attrs，可以只连接 standby 节点进行报表查询。另外，推荐设置参数 connect_timeout，如果某个节点未能及时连接，那么可以快速尝试使用新的节点连接。

7.4　PostgreSQL 16 的系统层变化

PostgreSQL 16 的系统层变化主要体现在版本兼容性变化、系统函数变化、预置角色变化、配置参数变化、初始用户权限优化、成员角色权限变化、附加模块变化。

7.4.1　版本兼容性变化

在 PostgreSQL 16 中，有如下版本兼容性的变化需要注意。
- postgres 进程名称统一，彻底抹除 postmaster 进程信息。

postmaster 进程架构是多年前的古老设计，数据库安装目录一直使用如下符号链接保持向下兼容。

```
/opt/pg15/bin/postmaster -> postgres
```

从 PostgreSQL 16 开始，相关文档及后台目录彻底抹除了 postmaster 进程信息，统一使用 postgres 进程。

- 备库升主库，删除 promote_trigger_file 参数文件触发方式。

promote_trigger_file 参数文件触发备库升主库的方式是 PostgreSQL 12 引入的，用来代替 PostgreSQL 12 之前的版本中使用 recovery.conf 文件配置参数 trigger_file 的方式。同时，PostgreSQL 12 提供了更便捷的 pg_promote 函数，或 pg_ctl promote 命令用于备库升主库。在 promote_trigger_file 参数文件触发方式下，流复制节点每隔 5s 会检测是否存在该文件，废弃这种方式可以降低能耗。

- 参数 vacuum_defer_cleanup_age 被删除。

参数 vacuum_defer_cleanup_age 作为降低流复制冲突的参数之一，多数用户很难合理设置其大小，并且使用该参数提高了事务 ID 边界计算的复杂性。此外，因设置该参数的值过大而引起数据损坏的情况时有发生。而在多数场景下，使用复制槽或参数 hot_standby_feedback 等也可以起到降低流复制冲突的作用。鉴于这些综合因素，PostgreSQL 16 删除了该参数。

- 参数 lc_collate 和 lc_ctype 被删除。

数据库在初始化时设置的参数 lc_collate 和 lc_ctype 是全局的参数，这两个参数是不可以修改的。在 PostgreSQL 16 中增加了对数据库级别单独设置本地化参数的功能。鉴于此，上面两个全局的本地化参数已经没有意义了，并且容易让用户对数据库级别的本地化设置产生困惑，为此 PostgreSQL 16 删除了上面两个全局的本地化参数。

- 提示优化器考虑使用并行特性：将参数 force_parallel_mode 重命名为 debug_parallel_query。

很多用户会认为 PostgreSQL 16 之前的版本中的参数 force_parallel_mode 是开启并行查询的开关，这种想法其实并不准确，参数 force_parallel_mode 只是确保优化器会考虑使用并行特性，而最终是否使用并行特性还会受成本评估的影响。使用该参数不会强迫优化器使用并行特性。为了消除用户对此产生的误解，PostgreSQL 16 将参数 force_parallel_mode 重命名为 debug_parallel_query。

- 参数 archive_library 与 archive_command 不能同时使用。

PostgreSQL 15 配置 WAL 文件归档，引入了一种新的配置方式，可以配置参数 archive_library，并且参数 archive_library 与 archive_command 允许同时使用，而参数 archive_library 的配置会优先生效。不允许在 PostgreSQL 16 中同时使用参数 archive_library 与 archive_command，否则数据库日志会出现错误提示。

- 非 ASCII 字符转换规则变更。

在数据库中部分参数的值只能被配置为 ASCII 字符。例如，在 PostgreSQL 16 之前的版本中，如果参数 application_name 和 cluster_name 的值设置了非 ASCII 字符，那么会显示为问号，代码如下。

```
postgres=# set application_name to 'pg15客户端';
SET
postgres=# show application_name;
```

```
application_name
-----------------
 pg15?????????
(1 row)
```

可以发现，在参数 application_name 中非 ASCII 字符"客户端"按照 UTF-8 编码字节数显示为 9 个问号。

而在 PostgreSQL 16 中，则以十六进制形式显示，代码如下。

```
postgres=# set application_name to 'pg16客户端';
SET
postgres=# show application_name;
           application_name
----------------------------------------
 pg16\xe5\xae\xa2\xe6\x88\xb7\xe7\xab\xaf
(1 row)
```

7.4.2 系统函数变化

PostgreSQL 16 中系统函数的变化主要如下。

新引入了 1 个 SQL 标准 2023 函数

- any_value：对分组后的数据去重，只返回单行。

新引入了 1 个常量函数

- SYSTEM_USER：SQL 保留关键字实现的常量函数，可以返回用户名及认证方式。

新引入了 3 个数学函数

- random_normal：标准正态分布数学函数。
- erf：与标准正态分布数学函数相关的误差函数。
- erfc：与标准正态分布数学函数相关的互补误差函数。

新引入了 2 个数组函数

- array_shuffle：对数组进行随机排序。
- array_sample：对数组返回多个抽样值。

新引入了 2 个日期函数

- date_add：对带时区的时间类型进行时区加法运算。
- date_subtract：对带时区的时间类型进行时区减法运算。

新引入了 2 个系统信息函数

- pg_input_is_valid：检测数据类型是否有效。
- pg_input_error_info：如果 pg_input_is_valid 函数检测输入值与数据类型不匹配，那么可以使用 pg_input_error_info 函数查看错误信息。

新引入了 3 个监控统计函数

- pg_stat_get_io：pg_stat_io 视图的内部实现。
- pg_stat_get_backend_subxact：返回后端子事务状态相关信息。
- pg_stat_get_lastscan：last_idx_scan 字段的内部实现。

新引入了 2 个系统管理函数

- pg_split_walfile_name：根据 WAL 文件名获取其序号及时间线。
- pg_log_standby_snapshot：捕获进行中事务的快照并将其存储到 WAL 文件中以支持备库逻辑解码。

其他变化

- array_agg 及 string_agg：支持并行操作。
- pg_create_logical_replication_slot：可以在备库中执行。
- xmlserialize：新增 indent 选项，用于控制是否进行格式化显示。
- pg_read_file 及 pg_read_binary_file：新增文件中不存在的检测选项，用于避免文件读取错误。

7.4.3 预置角色变化

PostgreSQL 16 新引入了两个预置角色，分别是 pg_create_subscription、pg_use_reserved_connections。

pg_create_subscription 允许执行 CREATE SUBSCRIPTION 命令而不需要具有超级用户权限。pg_use_reserved_connections 允许使用参数 reserved_connections 设置的连接数来执行一些后台运维及管理操作而无须担心出现因数据库连接被应用程序占满而无法正常工作的问题。

7.4.4 配置参数变化

PostgreSQL 16 中的参数变化主要如下：重命名了 1 个配置参数，新引入了 12 个配置参数（见表 7-1），并废弃了几个配置参数（参考 7.4.1 节内容）。

表 7-1 PostgreSQL 16 新引入及重命名的配置参数

参数名称	描述
debug_parallel_query	允许优化器考虑使用并行特性而不是强制使用并行特性。此参数名称是对参数 force_parallel_mode 的重命名
enable_presorted_aggregate	允许优化器对带 ORDER BY 子句或 DISTINCT 子句的聚合操作通过预排序来减少显式排序的代价消耗

续表

参数名称	描述
scram_iterations	可以控制使用 SCRAM-SHA-256 方式加密的迭代次数，以提升安全性
createrole_self_grant	允许在使用普通用户权限创建用户时，使用 set 选项的值切换成被创建的用户，或使用 inherit 选项的值继承被创建用户的全部权限
reserved_connections	允许拥有 pg_use_reserved_connections 的用户设置预留的连接数
vacuum_buffer_usage_limit	控制 VACUUM 操作使用 ring buffer 内存的大小
max_parallel_apply_workers_per_subscription	控制订阅端并行应用大事务的工作进程数
icu_validation_level	设置 ICU 排序规则验证的日志级别
gss_accept_delegation	控制是否允许使用 GSSAPI 进行代理认证
send_abort_for_crash	发送 SIGABRT 信号给子进程，而不是使用以前的 SIGQUIT 信号
send_abort_for_kill	发送 SIGABRT 信号给子进程，而不是使用以前的 SIGKILL 信号
debug_logical_replication_streaming	控制逻辑解码日志的时机，使用 buffered 选项的值需要等待逻辑解码的缓存填满后才能解析，使用 immediate 选项的值可以实时解析本参数主要用于开发调试
debug_io_direct	允许对操作系统内核使用 O_DIRECT 特性进行功能测试。打开本参数，数据库将不能使用操作系统的预取功能。由于数据库自身还未实现预取功能，会损耗性能，因此不推荐打开本参数

7.4.5 初始用户权限优化

在 PostgreSQL 16 中，使用 initdb 工具初始化数据库时生成的初始用户不允许移除超级用户权限，代码如下。

```
postgres=# alter user postgres nosuperuser;
ERROR:  permission denied to alter role
DETAIL:  The bootstrap user must have the SUPERUSER attribute.
```

上面的代码中对默认的初始用户 postgres 移除超级用户权限导致提示错误，这是因为初始用户仍然是许多重要系统对象的宿主，即便它没有超级用户权限也可以通过修改 pg_catalog 模式下的系统表来获得超级用户权限。因此，在 PostgreSQL 16 之前的版本中允许移除超级用户权限并不符合安全预期。

7.4.6 成员角色权限变化

在 PostgreSQL 15 中，超级用户使用 GRANT 命令进行成员角色权限操作，代码如下。

```
grant b to a granted by c;
```

上面的操作设置 b 用户为 a 用户的成员角色，并指定赋权者为 c 用户，实际操作由默认的 postgres 用户完成。从 pg_auth_members 系统表中也可以看到该操作记录。有人可能会好奇 c 用户是如何进行赋权操作的，这是因为 c 用户只有是 b 用户的成员才能进行操作，而实际上 c 用户并不是 b 用户的成员。

在 PostgreSQL 16 中，超级用户执行上面的操作会失败，代码如下。

```
postgres=# grant b to a granted by c;
ERROR:  permission denied to grant privileges as role "c"
DETAIL:  The grantor must have the ADMIN option on role "b".
```

在 PostgreSQL 16 中，只有具有成员角色并且使用 grant 命令操作时带上 admin 选项才能成功执行上面的操作，代码如下。

```
postgres=# grant b to c with admin true;
GRANT ROLE
postgres=# grant b to a granted by c;
GRANT ROLE
```

PostgreSQL 16 对 grant 命令的语法进行了扩展，新增了 with inherit true 选项和 with set 选项。当对新建用户赋权系统预置角色时，可以使用 with inherit true 选项，代码如下。

```
postgres=# grant pg_read_all_settings to a with inherit true;
GRANT ROLE
postgres=# \c - a
You are now connected to database "postgres" as user "a".
postgres=> show data_directory;
 data_directory
-----------------
 /opt/pgdata1600
(1 row)
```

在使用 with inherit true 选项赋予成员角色后，不需要显式切换到成员角色即可进行相关的操作。而在自定义的角色需要设置成员角色时，为了安全，需要显式切换到成员角色，此时可以设置 with set true 选项。

PostgreSQL 16 另外一个变化是 createrole 属性的权限，在 PostgreSQL 16 中创建用户时，使用 createrole 属性有如下变化。

- 自动成为新建用户的成员。
- 拥有新建用户的管理权限。
- 没有新建用户的继承权限。
- 没有新建用户的角色切换权限。

下面进行代码演示。

```
postgres=# create role admin login createrole;
CREATE ROLE
postgres=# \c - admin
You are now connected to database "postgres" as user "admin".
postgres=> create role user1 login;
CREATE ROLE
postgres=> \du admin
          List of roles
 Role name | Attributes  | Member of
-----------+-------------+-----------
 admin     | Create role | {user1}
postgres=> alter user user1 connection limit 10;
ALTER ROLE
```

上述第 1 行代码中被创建的 admin 用户带有 createrole 属性，从第 4 行代码中可以看到 admin 用户自动变为 user1 用户的成员，从第 5 行代码中可以看到 admin 用户可以对创建的 user1 用户设置连接限制，进行权限管理，并且只能对自己创建的角色或被赋予 admin 选项的角色进行管理。

在 PostgreSQL 16 中使用 createrole 属性创建的用户管理其他用户的能力受到了一定的限制，接下来的几行操作代码如下。

```
postgres=> \c - postgres
You are now connected to database "postgres" as user "postgres".
postgres=# create role admin2 LOGIN CREATEROLE;
CREATE ROLE
postgres=# \c - admin2
You are now connected to database "postgres" as user "admin2".
postgres=> alter user user1 connection limit -1;
ERROR:  permission denied to alter role
DETAIL:  Only roles with the CREATEROLE attribute and the ADMIN option on role
   "user1" may alter this role.
```

同样，拥有 createrole 属性的 admin2 用户并不能对 admin 用户创建的 user1 用户进行权限管理。可以从 pg_auth_members 系统表中查看到如下数据。

```
postgres=> SELECT roleid::regrole,member::regrole,grantor::regrole, *
        FROM pg_auth_members
        WHERE roleid::regrole::text !~ '^pg_' \gx
-[ RECORD 1 ]--+---------
roleid         | user1
member         | admin
grantor        | postgres
oid            | 49336
roleid         | 49335
member         | 49334
grantor        | 10
admin_option   | t
inherit_option | f
set_option     | f
```

观察上面的数据可知，PostgreSQL 16 对 pg_auth_members 系统表新增了两个字段：inherit_option 字段和 set_option 字段。admin_option 字段的值为 t 表示 admin 用户可以对 user1 用户进行权限管理，inherit_option 字段的值为 f 表示 admin 用户没有继承 user1 用户的权限，set_option 字段的值为 f 表示 admin 用户没有切换到 user1 用户的权限。

可以发现，在新建用户后，默认只具有管理用户的权限而不能直接继承用户的权限，并且不能直接切换用户。

PostgreSQL 16 为了保持向下兼容性，新引入了参数 createrole_self_grant，参数 createrole_self_grant 的值可以被配置为 inherit、set，或二者的组合，并使用逗号分隔。

下面是参数 createrole_self_grant 的使用示例。

```
postgres=# \c - admin
You are now connected to database "postgres" as user "admin".
postgres=> set createrole_self_grant to 'set';
SET
postgres=> create user user2;
CREATE ROLE
postgres=> set role user2;
```

```
SET
postgres=> select roleid::regrole,member::regrole,grantor::regrole, *
          from pg_auth_members
          where roleid::regrole::text = 'user2'
          and grantor::regrole::text = 'admin' \gx
-[ RECORD 1 ]--+------
roleid         | user2
member         | admin
grantor        | admin
oid            | 49358
roleid         | 49356
member         | 49334
grantor        | 49334
admin_option   | f
inherit_option | f
set_option     | t
```

由于参数 createrole_self_grant 的值被设置为 set，使用 admin 用户创建的 user2 用户可以被 admin 用户使用 SET 命令切换，同时 pg_auth_members 系统表的 set_option 字段的值也变为了 t。

7.4.7　附加模块变化

PostgreSQL 16 中附加模块的变化具体如下。
- pg_stat_statements 模块新增 DDL 语句的归一化监控功能。
- pg_walinspect 模块可以获取全页写信息。
- postgres_fdw 模块新增 parallel_abort 选项，对外部表提供了异步中止事务的能力。
- postgres_fdw 模块新增 analyze_sampling 选项，允许使用抽样功能收集外部表的统计信息。
- auto_explain 模块新增输出查询 ID 的功能。
- auto_explain 模块新增参数 log_parameter_max_length，可以控制输出日志的宽度。

7.5　本章小结

PostgreSQL 16 的性能的提升主要包括引入并行 HASH FULL JOIN 操作和 HASH RIGHT JOIN 操作，以及 string_agg 函数和 array_agg 函数支持并行聚合等。

PostgreSQL 16 的运维管理优化主要包括引入 pg_stat_io 视图，重点介绍该视图的 3 个大的标签维度及 8 个小的指标维度。此外，VACUUM 命令及 VACUUMDB 工具在功能方面有了大量的增强，包括引入内存可控的参数 vacuum_buffer_usage_limit，以及快速清理 TOAST 表等。

PostgreSQL 16 的开发易用性也有了较大的提升，比较突出的功能是逻辑复制功能，支持 standby 节点逻辑解码，以及解决循环复制。此外，SQL 标准 2023 支持非十进制的整型常量，以及下画线虚拟分组等。对 SQL/JSON 函数的功能进行了增强等。

PostgreSQL 16 的系统层有了一些变化，主要包括版本兼容性变化、系统函数变化、预置角色变化、配置参数变化等。

反侵权盗版声明

电子工业出版社依法对本作品享有专有出版权。任何未经权利人书面许可，复制、销售或通过信息网络传播本作品的行为；歪曲、篡改、剽窃本作品的行为，均违反《中华人民共和国著作权法》，其行为人应承担相应的民事责任和行政责任，构成犯罪的，将被依法追究刑事责任。

为了维护市场秩序，保护权利人的合法权益，我社将依法查处和打击侵权盗版的单位和个人。欢迎社会各界人士积极举报侵权盗版行为，本社将奖励举报有功人员，并保证举报人的信息不被泄露。

举报电话：（010）88254396；（010）88258888
传　　真：（010）88254397
E-mail：dbqq@phei.com.cn
通信地址：北京市万寿路173信箱
　　　　　电子工业出版社总编办公室
邮　　编：100036